Developments in Clay Science, 2

APPLIED CLAY MINERALOGY

Occurrences, Processing and Application of Kaolins, Bentonites, Palygorskite-Sepiolite, and Common Clays

Developments in Clay Science, 2

APPLIED CLAY MINERALOGY

Occurrences, Processing and Application of Kaolins, Bentonites, Palygorskite-Sepiolite, and Common Clays

HAYDN H. MURRAY

Professor Emeritus
Department of Geological Sciences
Indiana University
Bloomington, Indiana, U.S.A.

ELSEVIER

Amsterdam • Boston • Heidelberg • London • New York • Oxford
Paris • San Diego • San Francisco • Singapore • Sydney • Tokyo

Elsevier
Radarweg 29, PO Box 211, 1000 AE Amsterdam, The Netherlands
The Boulevard, Langford Lane, Kidlington, Oxford OX5 1GB, UK

First edition 2007

Library of Congress Cataloging-in-Publication Data
A catalogue record for this book is available from the British Library of Congress

British Library Cataloguing in Publication Data
A catalogue record for this book is available from the British Library

ISBN-13: 978-0-444-51701-2
ISBN-10: 0-444-51701-4
ISSN: 1572-4352

Transferred to Digital Printing 2007

07 08 09 10 11 10 9 8 7 6 5 4 3 2 1

CONTENTS

PREFACE

The author has had a career which involved academic teaching and research in the areas of clay mineralogy, sedimentology, and geology of industrial minerals; clay mineralogist for the Indiana Geological Survey; and 17 years in industry with Georgia Kaolin Company, a clay company with interests in kaolins and bentonites. At Georgia Kaolin Company, he had positions as Director of Research, Manager of Operations, Vice President of Operations, and Executive Vice President and Chief Operating Officer. Georgia Kaolin Company mined and processed kaolins in Georgia, sodium bentonites in Wyoming, calcium bentonites in Texas, and halloysite in New Zealand.

In recent years, he has been associated with companies that mined, processed, and marketed palygorskite in South Georgia and North Florida, palygorskite in Anhui and Jiangsu Provinces in China, ball clays from Western Tennessee, and kaolins from the Lower Amazon region in Brazil. From 1970 to 1981, he chaired a project sponsored by UNESCO to study the genesis of kaolins. This project involved an annual conference and field trips to visit and evaluate kaolin deposits in the United States, Europe, and Asia. Also, as a consultant, he has evaluated kaolin deposits in Argentina, Australia, Brazil, China, Egypt, Indonesia, Japan, Mexico, South Africa, Spain, and Venezuela. In addition, he has evaluated bentonite deposits in Argentina, Egypt, England, Algeria, Germany, and Chile plus a palygorskite deposit in Senegal in West Africa.

During his years in industry, he became interested in the many applications of clays and particularly the relationship between the structure, composition, and physical and chemical properties of the clay minerals and how these were related to their industrial applications. In this book, the structure and composition of the clay minerals, the geology and locations of the more important clay deposits, the mining and processing, and the many applications are discussed. In the appendices, the more important laboratory tests and procedures for evaluating kaolin and ball clays, bentonites, and palygorskite-sepiolite are described.

The author acknowledges with grateful thanks the contributions of his many graduate students including Cliff Ambers, Wayne M. Bundy, Thomas Dombrowski, Jessica Elzea-Kogel, Jack L. Harrison, Colin C. Harvey, Karan S. Keith, Roland Merkl, William F. Moll, Robert J. Pruett, Tim Salter, John M. Smith, Andy Thomas, Thomas Toth, Sue Weng, Jun Yuan, and Huitang Zhou. Also, thanks to his associates in Industry, Academia, and Government including Wayne M. Bundy, Robert F. Conley, William P. Hettinger, Jr., Fred Heivilin, Joe Iannicelli, Walter Keller, Sam Patterson, William Moll, John B. Patton, Joseph Shi, John M. Smith, Sam Smith, Paul Thiele, and especially to my mentor in graduate school and during my early career, Ralph E. Grim.

Also, I express my appreciation to my secretary, DeAnn Reinhart, for the many hours spent in typing and proofing the manuscript and to Kim Sowder and Barb Hill for their excellent help in preparing the photos and figures.

This book is dedicated to my wife, Juanita, for her patience and support in all my world travels and in writing this book.

Chapter 1

INTRODUCTION

Clays and clay minerals are very important industrial minerals. There are well over one hundred documented industrial applications of clay materials. Clays are utilized in the process industries, in agricultural applications, in engineering and construction applications, in environmental remediations, in geology, and in many other miscellaneous applications. This book is an assimilation of the major and minor uses of clays and clay minerals and explains why an understanding of the structure and physical and chemical attributes of the individual clay minerals are so important. Clay is an abundant raw material which has an amazing variety of uses and properties that are largely dependent on their mineral structure and composition. Other than the clay structure and composition, there are several additional factors which are important in determining the properties and applications of a clay. These are the non-clay mineral composition, the presence of organic material, the type and amount of exchangeable ions and soluble salts, and the texture (Grim, 1950).

First, the basic terms concerning clays and clay minerals must be defined. A clay material is any fine-grained, natural, earthy, argillaceous material (Grim, 1962). Clay is a rock term and is also used as a particle size term. The term clay has no genetic significance because it is used for residual weathering products, hydrothermally altered products, and sedimentary deposits. As a particle size term, the size fraction comprised of the smallest particles is called the clay fraction. The Wentworth scale defines the clay grade as finer than 4 μm (Wentworth, 1922), which is used by many engineers and soil scientists whereas clay scientists generally consider 2 μm as the upper limit of the clay size grade.

Grim (1968) summarized what he termed the clay mineral concept which stated that clays are composed essentially of a small group of extremely small crystalline particles of one or more members of a group of minerals that are commonly known as the clay minerals. The clay minerals are hydrous aluminum silicates and in some of these minerals, iron and magnesium substitute for the aluminum and in some there are alkaline and alkaline earth elements present as essential constituents as will be discussed in Chapter 2. The clay mineral groups are kaolin, smectite, palygorskite–sepiolite, which are sometimes referred to as hormites (Martin-Vivaldi and Robertson, 1971) (the term has not been accepted by the International Nomenclature Committee); illite, chlorite, and mixed-layered clays. The properties of these clays are very different which are related to their structure and composition (Murray, 2000a).

The clay mineral composition refers to the relative abundance and identity of the clay minerals present in a clay material. In some instances, very small amounts of certain clay minerals have a large impact on the physical properties. An example is a kaolin that has a small percentage of smectite present. This may alter the low and high shear viscosity detrimentally. Also, the degree of crystal perfection of the kaolinite present affects the physical properties of the kaolin. A well-ordered kaolinite will have different properties than a poorly ordered kaolinite (Murray and Lyons, 1956). The identity of all the clay minerals present in a clay material must be determined in order to evaluate the physical properties (Murray, 2000a).

The non-clay mineral composition is also important because in many cases the non-clay minerals can significantly affect the properties of a clay material. An example is the presence of a fine particle quartz in a kaolin which seriously affects the abrasiveness of the kaolin (Murray, 2000b).

Organic material in a clay affects the color and other properties. In some cases, the presence of organic material is advantageous as in ball clays, and in others, is detrimental because it affects the brightness and whiteness of kaolin clays. Special organic clad clays such as sodium montmorillonite are processed to become organophilic and/or hydrophobic for special applications (Jordan, 1949).

The exchangeable ions and soluble salts affect the physical properties of a clay material. A calcium montmorillonite has very different viscosity and gelling characteristics than a sodium montmorillonite (Hendricks, 1945). The presence of soluble salts can flocculate a clay which causes a problem in processing the clay.

The texture of a clay material refers to the particle size distribution of the constituents, the particle shape, the orientation of the particles with respect to each other, and the forces which bind the particles together. The particle size distribution and the particle shape are very important properties in kaolins and ball clays (Murray, 2000b). The orientation of the particles and the forces which bind them together can shed a great deal of information about the environment of deposition (Murray, 1976).

As pointed out by Grim (1988), prior to the 1920s, geologists making analyses of sediments listed the finest particles as clay with no identification of what this material actually was. There was no adequate analytical technique for identifying the ultra-fine particles making up the clay material. The first American geologist to specialize in the study of clays was Prof. Heinrich Ries of Cornell University. He studied the clay resources of many of the eastern states by describing their ceramic properties (Ries, 1908). In the middle and late 1920s, X-ray diffraction began to be used to identify the clay minerals. Several scientists in the United States and Europe published studies of clays using X-ray diffraction to positively identify the clay materials (Hadding, 1923; Rinne, 1924; Hendricks and Fry, 1930; Ross and Kerr, 1930, 1931).

At the present time, much more sophisticated analytical equipment is available to identify and quantify the specific clay minerals present in a sample. Some of the more important analytical techniques that are used include X-ray diffraction, electron microscopy, infrared spectroscopy, and differential thermal analysis. Several books and articles have been published describing these techniques, a few of which are Brindley and Brown (1980), Moore and Reynolds (1997), Mackenzie (1970, 1972), Van der Marel and Beutelspacher (1976), and Sudo, Shimoda, Yutsumoto, and Aita (1981).

The technological properties of clay materials are largely dependent on a number of factors. As will be pointed out in this book, the physical and chemical properties of a clay are related to its structure and composition and on the type of processing used to beneficiate the clay product. The structure and composition of kaolins, smectites, and palygorskite–sepiolite are very different even though the fundamental building blocks, i.e. the tetrahedral and octahedral sheets, are similar. However, the arrangement and composition of the octahedral and tetrahedral sheets account for major and minor differences in the physical and chemical properties that control the applications of a particular clay mineral. Also important is the type and amount of non-clay minerals that are present. Non-clay minerals commonly associated with the clay minerals include quartz, feldspar, mica, calcite, dolomite, opal C-T, and minor amounts of heavy and trace minerals such as ilmenite, rutile, brookite, anatase, leucoxene, sphene, tourmaline, zircon, kyanite, goethite, hematite, magnetite, garnet, augite, florencite, apatite, andalusite, and barite.

There are several societies and groups that are specifically devoted to clay science and some publish journals, monographs, and special papers. Also, there are other societies and magazines that have divisions or sections in which clay papers are presented and/or published. The major societies and groups that are currently active in clay science are: The Clay Minerals Society in the United States, European Clay Group, which includes those from Great Britain, France, Germany, Spain, Portugal, Italy, Scandinavia, Poland, Czech Republic, and Slovenia; The Clay Science Society of Japan and Association Internationale pour l'Etude des Argiles (AIPEA). The Czech National Clay Group sponsors meetings periodically and publishes the proceedings.

The Clay Minerals Society hosts an annual conference and publishes the journal *Clays and Clay Minerals* and also special publications and workshop presentations. The European Clay Groups hold a Euroclay Conference every 2 years and publish the journal *Clay Minerals*. The Clay Science Society of Japan sponsors an annual conference and publishes the journal *Clay Science*. The AIPEA sponsors the International Clay Minerals Conference every 4 years and publishes the proceedings of each conference. The journal *Applied Clay Science* is published by Elsevier. Other organizations and publications which may contain articles on clays are the American Ceramic Society (annual meetings and bulletin), The Society for Mining, Metallurgy, and Exploration, Inc. (annual meetings, preprints, books, *Mining Engineering* magazine, and transactions), and *Industrial Minerals* magazine.

Many other individual countries and regions have active clay mineral groups including Argentina, Australia, Brazil, India, and Israel.

REFERENCES

Brindley, G.W. and Brown, G. (1980). *Crystal Structures of Clay Minerals and their X-Ray Identification*. Mineralogical Society Monograph No. 5, London, 495pp.

Grim, R.E. (1950) Modern concepts of clay materials. *J. Geol.*, 50, 225–275.

Grim, R.E. (1962) *Applied Clay Mineralogy*. McGraw-Hill, New York, 422pp.

Grim, R.E. (1968) *Clay Mineralogy*, 2nd Edition. McGraw-Hill, New York, 596pp.

Grim, R.E. (1988) The history of the development of clay mineralogy. *Clay. Clay Miner.*, 36, 97–101.

Hadding, A. (1923) Eine Rötgenographische Methode Kristalline and Kryptokristalline Substanzen Zu Identifizieren. *Z. Kristallogr.*, 58, 108–122.

Hendricks, S.B. (1945) Base exchange in the crystalline silicates. *Ind. Eng. Chem.*, 37, 625–630.

Hendricks, S.B. and Fry, W.H. (1930) The results of X-ray and microscopic examination of soil colloids. *Soil Sci.*, 29, 457–478.

Jordan, J.W. (1949) Organophilic bentonites. *J. Phys. Colloid Chem.*, 53, 294–306.

Mackenzie, R.C. (1970) *Differential Thermal Analysis of Clays. Vol. 1: Fundamental Aspects*. Academic Press, New York.

Mackenzie, R.C. (1972) *Differential Thermal Analysis of Clays. Vol. 2: Applications*. Academic Press, New York.

Martin-Vivaldi, J.I. and Robertson, R.H.S. (1971). Palygorskite and sepiolite (the hormites). Chapter in *Electron Optical Investigation of Clays.* Gard, J.A., ed. Mineralogical Society Monograph No. 31, London, pp. 255–275.

Moore, D.M. and Reynolds, R.C. Jr. (1997) *X-Ray Diffraction and the Identification and Analysis of Clay Minerals*, 2nd Edition. Oxford University Press, Oxford and New York, 378pp.

Murray, H.H. (1976) The Georgia sedimentary kaolins. *Proceedings of the 7th International Kaolin Symposium*. Shimoda, S. ed. University of Tokyo, Tokyo, pp. 114–125.

Murray, H.H. (2000a) *Clays. Ullmann's Encyclopedia of Industrial Chemistry*, 6th Edition. Wiley-VCH Verlag GmBH, Weinheim, Germany, 30pp.

Murray, H.H. (2000b) Traditional and new applications for kaolin, smectite and palygorskite: a general overview. *Appl. Clay Sci.*, 17, 207–221.

Murray, H.H. and Lyons, S.C. (1956) Correlation of paper-coating quality with degree of crystal perfection of kaolinite. *Clay. Clay Miner.*, 456, 31–40.

Ries, H. (1908) *Clays—their Occurrence, Properties and Uses*, 2nd Edition. John Wiley and Sons, Inc, New York, 554pp.

Rinne, F. (1924) Rötgenographische Untersuchungen an Einigen Feinzerteilten Mineralien, Kunspredukten und Dichten Gesteinem. *Z. Kristallogr.*, 60, 55–69.

Ross, C.S. and Kerr, P.F. (1930) The clay minerals and their identity. *J. Sediment. Petrol.*, 1, 35–65.

Ross, C.S. and Kerr, P.F. (1931). *The Kaolin Minerals.* US Geological Survey, Professional Paper 165F, pp. 151–175.

Sudo, T., Shimoda, S., Yutsumoto, H., and Aita, S. (1981) *Electron Micrographs of Clay Minerals.* Elsevier Scientific Publishing Company, Amsterdam, Oxford, and New York, 203pp.

Van der Marel, H.W. and Beutelspacher, H. (1976) *Atlas of Infrared Spectroscopy of Clay Minerals and their Admixtures.* Elsevier Scientific Publishing Company, Amsterdam, Oxford, and New York, 396pp.

Wentworth, C.K. (1922) A scale of grade and class terms for clastic sediments. *J. Geol.*, 30, 377–392.

SOME ADDITIONAL REFERENCE BOOKS ON CLAYS AND CLAY MINERALS

Card, J.A. (1971) *The Electron-Optical Investigation of Clays.* Mineralogical Society, London, 383pp.

Chamley, H. (1989) *Clay Sedimentology*. Springer-Verlag, New York, 623pp.

Chukrov, F.V. (1955) *Colloids in the Earth's Crust.* Academy of Science, USSR, 672pp.

Fripiat, J.J. (1982) *Advanced Techniques for Clay Mineral Analysis. Developments in Sedimentology*, Vol. 34. Elsevier, Amsterdam, 235pp.

Grim, R.E. and Guven, N. (1978) *Bentonites—Geology, Mineralogy, Properties and Uses. Developments in Sedimentology*, Vol. 24. Elsevier, Amsterdam, 256pp.

Grimshaw, R.W. (1971) *The Chemistry and Physics of Clays.* Wiley-Interscience, New York, 1024pp.

Marshall, C.G. (1949) *The Colloid Chemistry of the Silicate Minerals.* Academic Press, New York, 146pp.

Millot, G. (1970) *Geology of Clays.* Springer-Verlag, New York, 425pp.

Murray, H.H., Bundy, W.M., and Harvey, C.C. (1993). *Kaolin Genesis and Utilization.* Special Publication No. 1. Clay Minerals Society, Boulder, CO, 341pp.

Newman, A.C.D. (Ed.), (1987). *Chemistry of Clays and Clay Minerals.* Mineralogical Society Monograph No. 6, London 480pp.

Robertson, R.H.S. (1986) *Fuller's Earth—A History of Calcium Montmorillonite.* Volturna Press, Hythe, Kent, UK, 421pp.

Siddiqui, M.K.H. (1968) *Bleaching Earths.* Pergamon Press, New York, 86pp.

Sudo, T. and Shinoda, S. (1978) *Clays and Clay Minerals of Japan.* Elsevier, New York, 326pp.

Theng, B.K.G. (1979) *Formation and Properties of Clay–Polymer Complexes.* Elsevier, Amsterdam, 362pp.

Van Olphen, H. (1977) *An Introduction to Clay Colloid Chemistry*, 2nd Edition. John Wiley and Sons, New York, 318pp.

Velde, B. (1977) *Clay and Clay Minerals in Natural and Synthetic Systems. Developments in Sedimentology*, Vol. 21. Elsevier, Amsterdam, 218pp.

Velde, B. (1985) *Clay Minerals—A Physico-Chemical Explanation of their Occurrence. Developments in Sedimentology*, Vol. 40. Elsevier, Amsterdam, 427pp.

Weaver, C.E. (1989) *Clays, Muds and Shales. Developments in Sedimentology*, Vol. 44. Elsevier, Amsterdam, 819pp.

Weaver, C.E. and Pollard, L.D. (1973) *The Chemistry of Clay Minerals. Developments in Sedimentology*, Vol. 15. Elsevier, Amsterdam, 213pp.

Wilson, M.J. (1994) *Clay Mineralogy—Spectroscopic and Chemical Determination Methods.* Chapman and Hall, London, 367pp.

Zvyagin, B.B. (1967) *Electron-Diffraction Analysis of Clay Mineral Structures.* Plenum Press, New York, 364pp.

Chapter 2

STRUCTURE AND COMPOSITION OF THE CLAY MINERALS AND THEIR PHYSICAL AND CHEMICAL PROPERTIES

In this chapter, a general review of the structure and composition of the various clay minerals are given. Those who are interested in more detailed discussions of the structures should consult Guven (1988), Jones and Galan (1988), Bailey (1980, 1988, 1993), and Moore and Reynolds (1997). The physical and chemical properties of a particular clay mineral are dependent on its structure and composition.

A useful classification of the clay minerals (Table 1) was proposed and used by Grim in his book (1968), which is a basis for outlining the nomenclature and differences between the various clay minerals.

The atomic structure of the clay minerals consists of two basic units, an octahedral sheet and a tetrahedral sheet. The octahedral sheet is comprised of closely packed oxygens and hydroxyls in which aluminum, iron, and magnesium atoms are arranged in octahedral coordination (Fig. 1). When aluminum with a positive valence of three is the cation present in the octahedral sheet, only two-thirds of the possible positions are filled in order to balance the charges. When only two-thirds of the positions are filled, the mineral is termed dioctahedral. When magnesium with a positive charge of two is present, all three positions are filled to balance the structure and the mineral is termed trioctahedral.

The second structural unit is the silica tetrahedral layer in which the silicon atom is equidistant from four oxygens or possibly hydroxls arranged in the form of a tetrahedron with the silicon atom in the center. These tetrahedrons are arranged to form a hexagonal network repeated infinitely in two horizontal directions to form what is called the silica tetrahedral sheet (Fig. 2). The silica tetrahedral sheet and the octahedral sheet are joined by sharing the apical oxygens or hydroxyls to form what is termed the 1:1 clay mineral layer (e.g. kaolinite) or the 2:1 clay mineral layer (e.g. illite) as discussed in the following sections. The structure and composition of the major industrial clays, i.e. kaolins, smectites, and palygorskite–sepiolite, are very different even though they are each

Table 1. Classification of the clay minerals

I. Amorphous
   Allophane group

II. Crystalline
   A. Two-layer type (sheet structures composed of units of one layer of silica tetrahedrons and one layer of alumina octahedrons)
      1. Equidimensional
         Kaolinite group
         Kaolinite, dickite and nacrite
      2. Elongate
         Halloysite

   B. Three-layer types (sheet structures composed of two layers of silica tetrahedrons and one central dioctahedral or trioctahedral layer)
      1. Expanding lattice
         a. Equidimensional
            Smectite group
            Sodium montmorillonite, calcium montmorillonite, and beidellite
            Vermiculite
         b. Elongate
            Smectite
            Nontronite, saponite, hectorite
      2. Non-expanding lattice
         Illite group

   C. Regular mixed-layer types (ordered stacking of alternate layers of different types)
      Chlorite group

   D. Chain-structure types (hornblende-like chains of silica tetrahedrons linked together by octahedral groups of oxygens and hydroxyls containing Al and Mg atoms)
      Sepiolite
      Palygorskite (attapulgite)

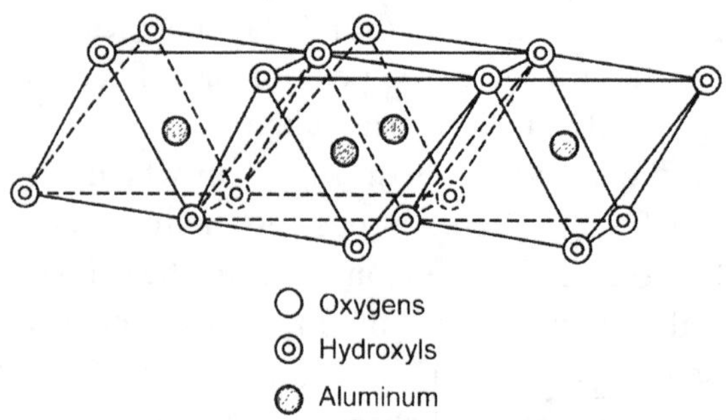

Fig. 1. Diagrammatic sketch of the octahedral sheet.

comprised of octahedral and tetrahedral sheets as their basic building blocks. The arrangement and composition of the octahedral and tetrahedral sheets account for most of the differences in their physical and chemical properties.

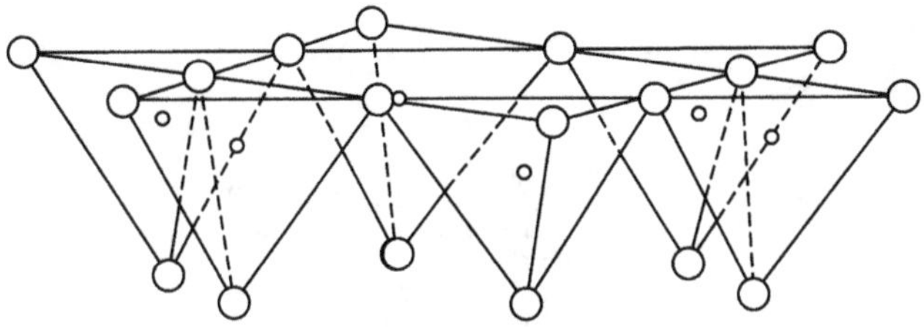

Fig. 2. Diagrammatic sketch of the tetrahedral sheet.

1. KAOLIN MINERALS

The basic kaolin mineral structure comprising the minerals kaolinite, dickite, nacrite, and halloysite is a layer of a single tetrahedral sheet and a single octahedral sheet. These two sheets are combined to form a unit in which the tips of the silica tetrahedrons are joined with the octahedral sheet. All of the apical oxygens of the silica tetrahedrons point in the same direction so that these oxygens and/or hydroxyls, which may be present to balance the charges, are shared by the silicons in the tetrahedral sheet and the aluminum in the octahedral sheet (Fig. 3). The structural formula for kaolinite is $Al_4Si_4O_{10}(OH)_8$ and the theoretical chemical composition is SiO_2, 46.54%; Al_2O_3, 39.50%; and H_2O, 13.96%.

Only two-thirds of the octahedral positions are filled by an aluminum atom. The aluminum atoms are surrounded by four oxygens and eight hydroxyls. The charge distribution in the kaolinite layer is shown in Table 2.

The charges in the kaolinite structure are balanced. The minerals of the kaolin group, kaolinite, dickite, nacrite, and halloysite consist of the so-called 1:1 layers of combined octahedral and tetrahedral sheets, which are continuous in the *a*- and *b*-axis directions and are stacked one above the other in the *c*-axis direction (Fig. 3). The differences in the kaolin minerals are the manner in which the unit layers are stacked above each other. The thickness of the unit layer is 7.13 Å.

In dickite, the unit cell consists of two unit layers and in nacrite, six unit layers. Halloysite occurs in two forms: one hydrated, in which there is a layer of water molecules between the layers, and one dehydrated. The hydrated form has a basal spacing of 10 Å (Fig. 4) and the dehydrated form, 7.2 Å. The shape of halloysite is elongate tubes (Fig. 5) whereas the shape of kaolinite is pseudo-hexagonal plates and stacks (Fig. 6). The International Nomenclature Committee has recommended the terms 7 Å

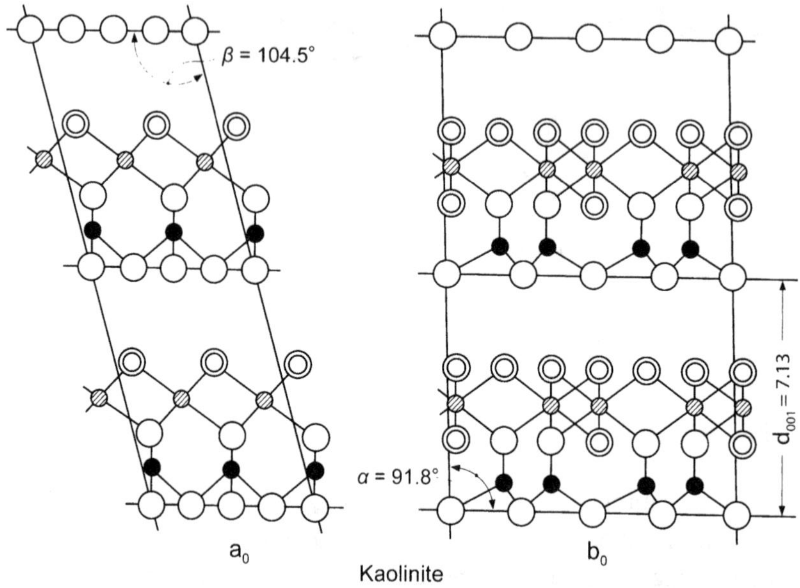

Fig. 3. Diagrammatic sketch of the structure of kaolinite.

Table 2. Charge distribution of the kaolinite layer

$6O^{2-}$	12^{-}
$4Si^{4+}$	16^{+}
$4O^{2-} + 2(OH)^{-}$	10^{-} (Layer shared by the tetrahedral and octahedral sheets)
$4Al^{3+}$	12^{+}
$6(OH)^{-}$	6^{-}

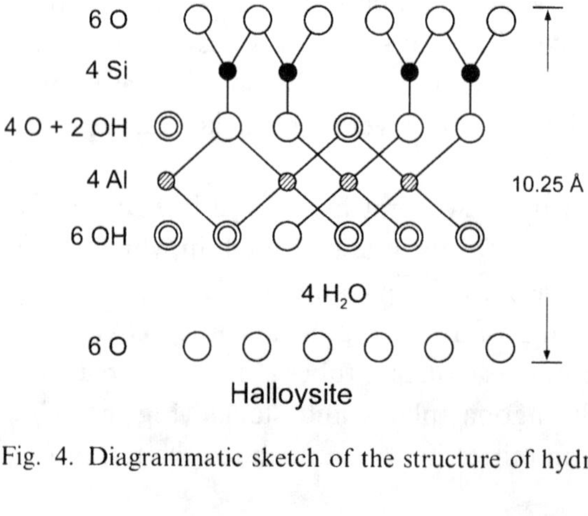

Fig. 4. Diagrammatic sketch of the structure of hydrated halloysite.

Fig. 5. Scanning electron micrograph of halloysite.

Fig. 6. Scanning electron micrograph of kaolinite.

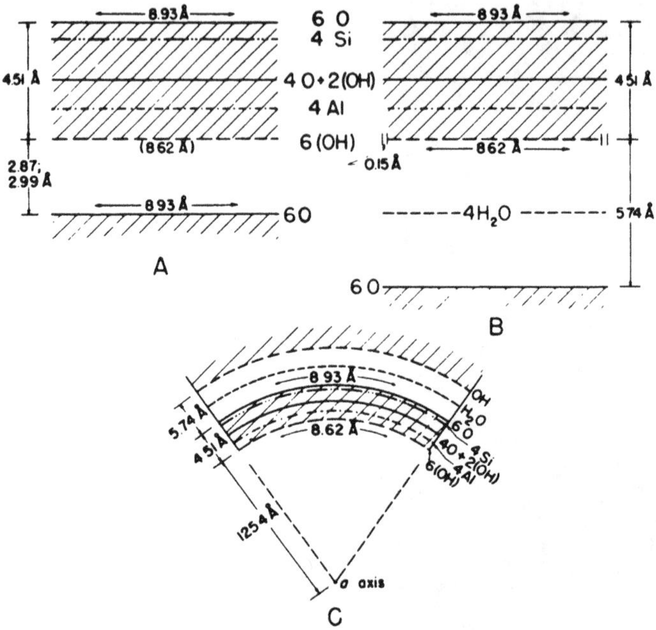

Fig. 7. Diagrammatic sketch of the structure of kaolinite and hydrated halloysite (Bates et al., 1950): (a) kaolinite; (b) hydrated halloysite; and (c) proposed cause for tubular shape of hydrated halloysite.

halloysite and 10 Å halloysite to designate the two forms. The elongate tubular form according to Bates et al. (1950) is made up of overlapping curved sheets of the kaolinite type. The curvature develops in 10 Å halloysite because of the irregular stacking of the layers and the interlayer of water molecules, which cause a weak bond between the layers. The tendency to curve is caused by a slight difference in dimension of the silicon tetrahedral sheet and the alumina octahedral sheet (Fig. 7).

2. SMECTITE MINERALS

The major smectite minerals are sodium montmorillonite, calcium montmorillonite, saponite (magnesium montmorillonite), nontronite (iron montmorillonite), hectorite (lithium montmorillonite), and beidellite (aluminum montmorillonite). Smectite minerals are composed of two silica tetrahedral sheets with a central octahedral sheet and are designated as a 2:1 layer mineral (Fig. 8). Water molecules and cations occupy the space between the 2:1 layers.

The theoretical charge distribution in the smectite layer without considering substitutions in the structure is as shown in Table 3.

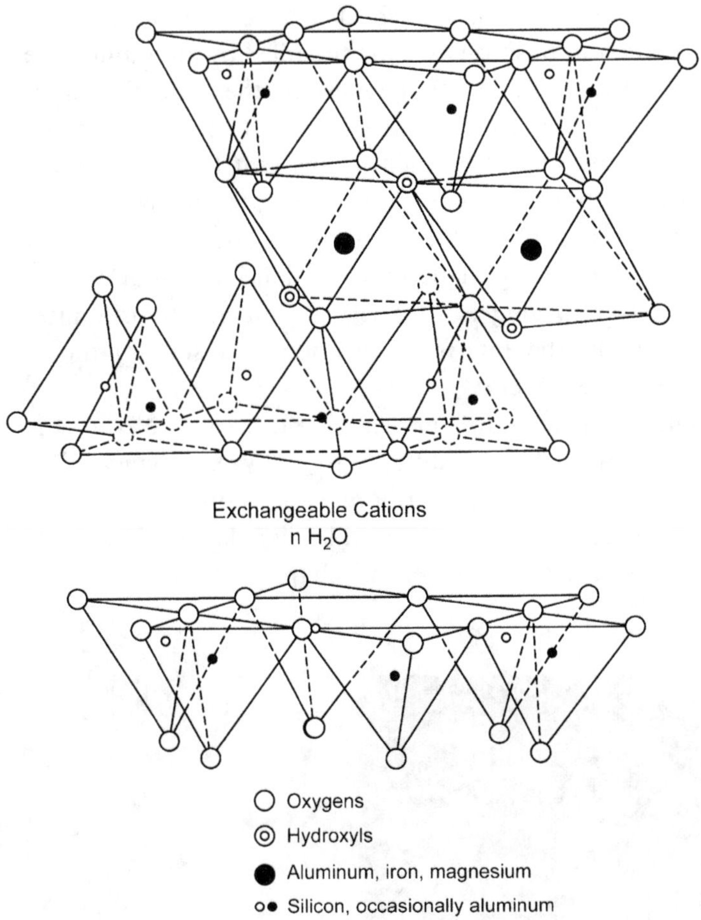

Fig. 8. Diagrammatic sketch of the structure of smectites.

Table 3. Charge distribution of the smectite layer

$6O^{2-}$	12^-
$4Si_4^+$	16^+
$4O^{2-}+2(OH)^-$	10^- (Plane common to tetrahedral and octahedral sheets)
$4Al^{3+}$	12^+
$4O^{2-}+2(OH)^-$	10^- (Plane common to tetrahedral and octahedral sheets)
$4Si^{4+}$	16^+
$6O^{2-}$	12^-

The theoretical formula is $(OH)_4Si_8Al_4O_{20} \cdot NH_2O$ (interlayer) and the theoretical composition without the interlayer material is SiO_2, 66.7%; Al_2O_3, 28.3%; and H_2O, 5%. However, in smectites, there is considerable substitution in the octahedral sheet and some in the tetrahedral sheet. In the tetrahedral sheet, there is substitution of aluminum for silicon up to

15% (Grim, 1968) and in the octahedral sheet, magnesium and iron for aluminum. If the octahedral positions are mainly filled by aluminum, the smectite mineral is beidellite; if filled by magnesium, the mineral is saponite; and if by iron, the mineral is nontronite. The most common smectite mineral is calcium montmorillonite, which means that the layer charge deficiency is balanced by the interlayer cation calcium and water. The basal spacing of the calcium montmorillonite is 14.2 Å. Sodium montmorillonite occurs when the charge deficiency is balanced by sodium ions and water and the basal spacing is 12.2 Å. Calcium montmorillonites have two water layers in the interlayer position and sodium montmorillonites have one water layer.

The smectite mineral particles are very small and because of this, the X-ray diffraction data are sometimes difficult to analyze. A typical electron micrograph of sodium montmorillonite is shown in Fig. 9. Smectites, and particularly sodium montmorillonites, have a high base exchange capacity as is described later in this chapter.

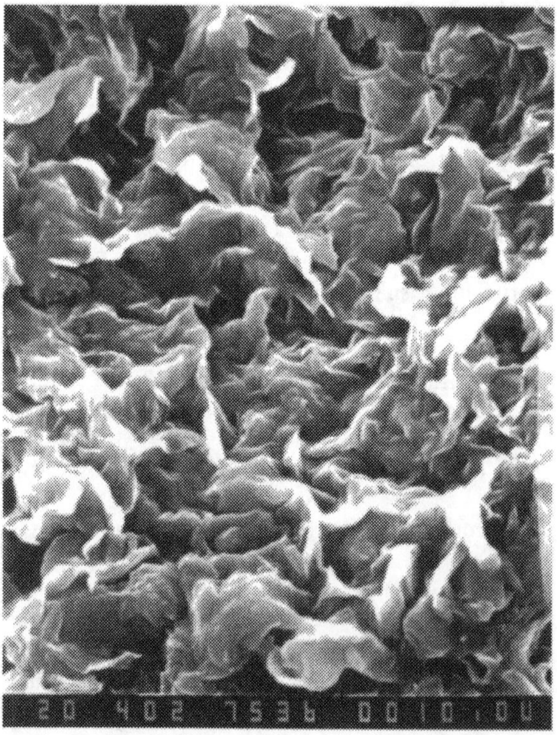

Fig. 9. Scanning electron micrograph of sodium montmorillonite from the Clay Spur Member of the Mowry Formation near Belle Fourche, SD.

3. ILLITE

Illite is a clay mineral mica, which was named by Grim et al. (1937). The structure is a 2:1 layer in which the interlayer cation is potassium (Fig. 10). The size, charge, and coordination number of potassium is such that it fits snugly in the hexagonal ring of oxygens of the adjacent silica tetrahedral sheets. This gives the structure a strong interlocking ionic bond which holds the individual layers together and prevents water molecules from occupying the interlayer position as it does in the smectites. A simple way of thinking about illite is that it is a potassium smectite.

Illite differs from well-crystallized muscovite in that there is less substitution of Al^{3+} for Si^{4+} in the tetrahedral sheet. In muscovite, one-fourth of the Si^{4+} is replaced by Al^{3+} whereas in illite only about one-sixth is replaced. Also, in the octahedral sheet, there may be some

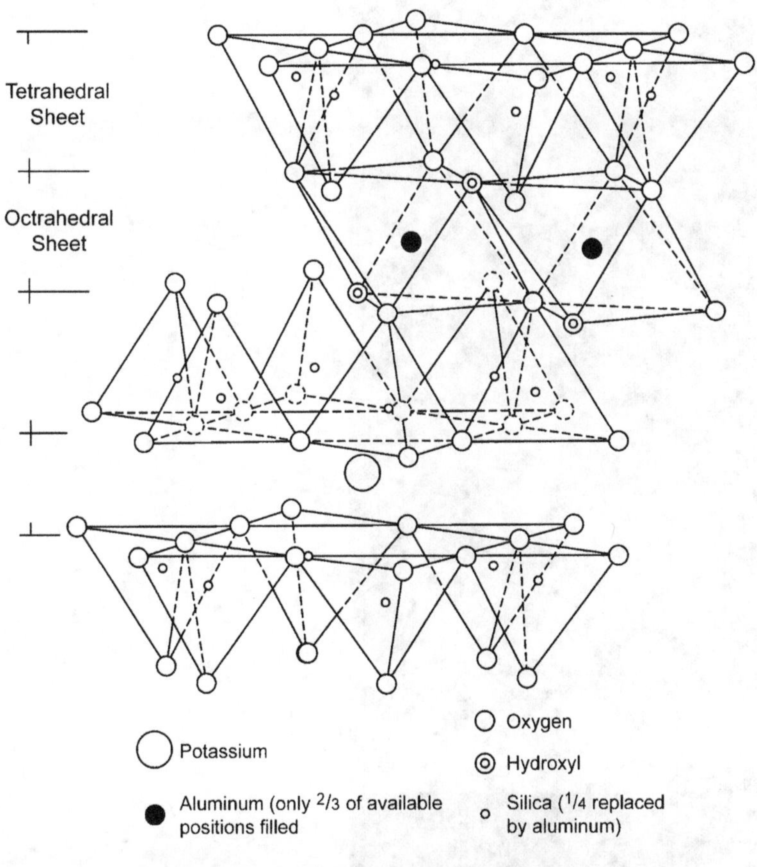

Fig. 10. Diagrammatic sketch of the structure of illite.

replacements of Al^{3+} by Mg^{2+} and Fe^{2+}. The basal spacing d(001) of illite is 10 Å. A more detailed discussion of the structure of illite and its variable composition can be found in Moore and Reynolds (1997). The charge deficiency, because of substitutions per unit cell layer, is about 1.30–1.50 for illite contrasted to 0.65 for smectite. The largest charge deficiency is in the tetrahedral sheet rather than in the octahedral sheet, which is opposite from smectite. For this reason and because of the fit, potassium bonds the layers in a fixed position so that water and other polar compounds cannot readily enter the interlayer position and also the potassium ion is not readily exchangeable. Fig. 11 is an electron micrograph of a Fithian illite. Fithian, Illinois is the location where Grim et al. (1937) described and named the clay mineral mica illite. Illite is commonly associated with many kaolins and smectites.

Fig. 11. Scanning electron micrograph of Fithian, Illinois illite.

4. CHLORITE

Chlorite is commonly present in shales and also in underclays associated with coal seams. Clay mineral chlorites differ from well-crystallized chlorites in that there is random stacking of the layers and also some hydration. Chlorite is a 2:1 layer mineral with an interlayer brucite sheet ($Mg(OH)_2$) (Fig. 12). There is quite a range of cation substitutions in chlorites, most commonly Mg^{2+}, Fe^{2+}, Al^{3+}, and Fe^{3+}. Those interested in a very detailed discussion of the structure of chlorite should consult Bailey (1988).

The composition of chlorite is generally shown as $(OH)_4(SiAl)_8(Mg\text{-}Fe)_6O_{20}$. The brucite-like sheet in the interlayer position has the general

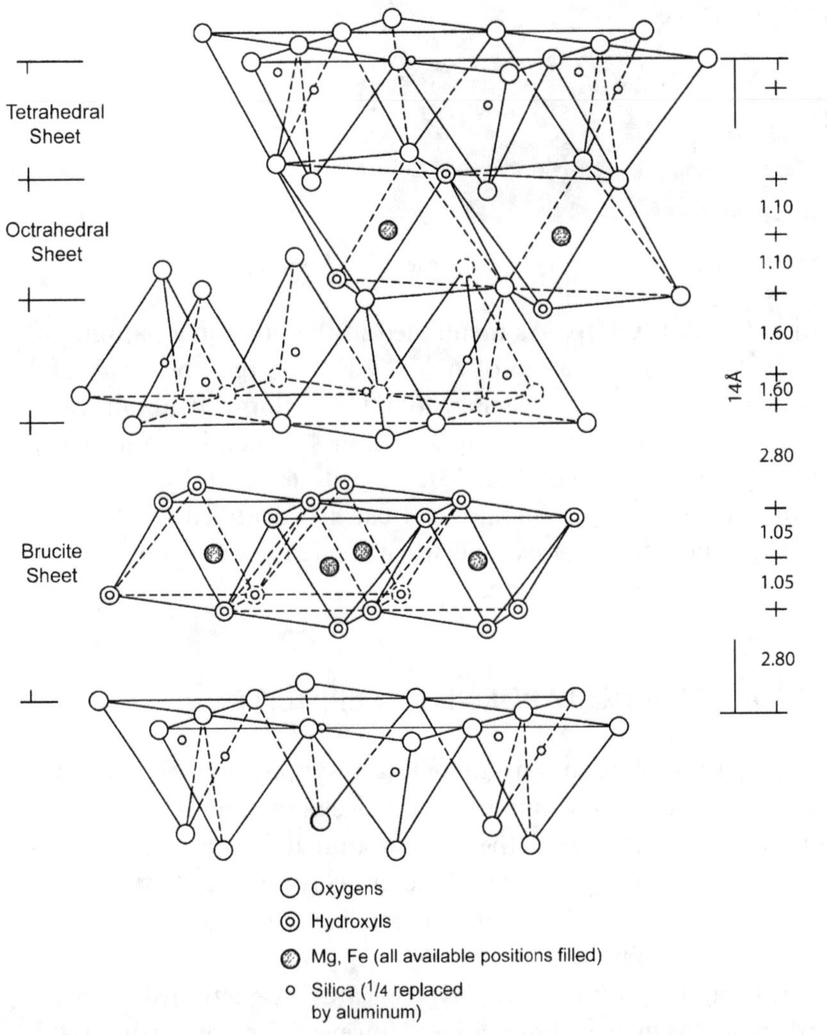

Fig. 12. Diagrammatic sketch of the structure of chlorite.

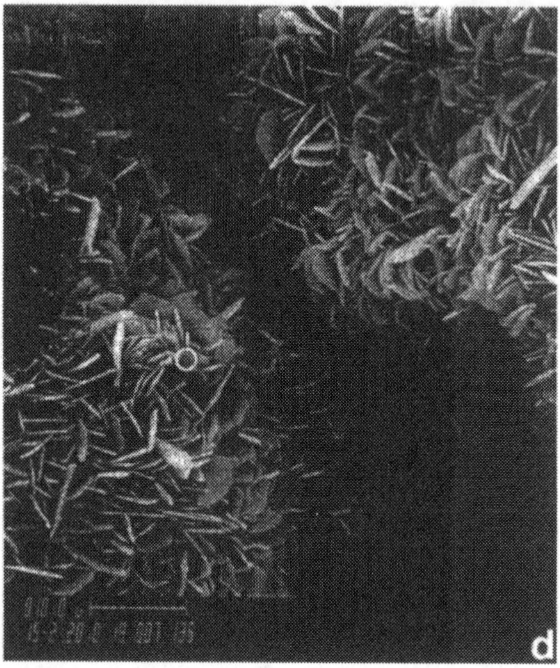

Fig. 13. Scanning electron micrograph of chlorite.

composition $(MgAl)_6(OH)_{12}$. As mentioned in the preceding paragraph, there is considerable substitution of Al^{3+} by Fe^{3+}, Mg^{2+} by Fe^{2+}, and of Si^{4+} by Al^{3+}. The basal spacing d(001) of chlorite is about 14 Å. Chlorite has been identified in many sandstones as coatings on quartz grains that appear as rosettes (Fig. 13). Chlorite is generally intimately intermixed with other clay minerals so it can be identified by the 14 Å basal spacing which does not expand when treated with ethylene glycol nor decrease to 10 Å upon heating.

5. PALYGORSKITE (ATTAPULGITE): SEPIOLITE

The terms palygorskite and attapulgite are synonymous, but the International Nomenclature Committee has declared that the preferred name is palygorskite. However, the term attapulgite is still used, particularly by those that mine, process, and use this clay mineral. In this book, the term palygorskite will be used, but readers should be aware that attapulgite is the same mineral.

Palygorskite and sepiolite are 2:1 layer silicates. The tetrahedral sheets are linked infinitely in two dimensions. However, they are structurally different from other clay minerals in that the octahedral sheets are continuous in only one dimension and the tetrahedral sheets are divided into ribbons by the periodic inversion of rows of tetrahedrons. The structures of palygorskite and sepiolite are shown in Fig. 14.

As shown in Fig. 14, the channels between ribbon strips are larger in sepiolite than in palygorskite. In palygorskite, the dimension of the channel is approximately 4 Å by 6 Å and in sepiolite, approximately 4 Å by 9.5 Å. Both of these clay minerals are magnesium silicates, but palygorskite has a higher alumina content. A general formula for palygorskite is $(OH_2)_4(OH_2)Mg_5Si_8O_{20} \cdot 4H_2O$. A general formula for sepiolite is $(OH_2)_4(OH)_4Mg_8Si_{12}O_{30} \cdot 8H_2O$.

As shown in Fig. 14, the *b*-axis in palygorskite is approximately 18 Å and in sepiolite it is about 27 Å. These two clay minerals contain two kinds of water, one coordinated to the octahedral cations and the other loosely bonded in the channels, which is termed zeolitic water. These channels may also contain exchangeable cations. Fig. 15 shows an electron micrograph of palygorskite. Both palygorskite and sepiolite are elongate in shape and often occur as bundles of elongate and lath-like particles. Usually, the sepiolite elongates are longer than palygorskite elongates (10–15 Å for sepiolite and >5 Å for palygorskite). The morphology of these two clay minerals is a most important physical attribute.

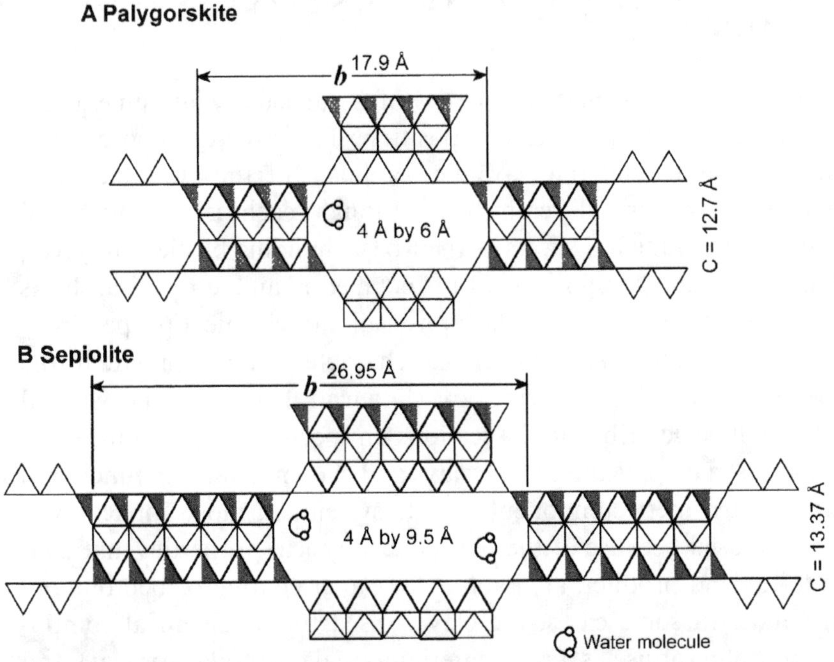

Fig. 14. Diagrammatic sketch of the structure of (a) palygorskite and (b) sepiolite.

Fig. 15. Scanning electron micrograph of palygorskite.

6. PHYSICAL AND CHEMICAL PROPERTIES OF CLAYS AND CLAY MINERALS

The physical and chemical properties of a particular clay mineral are dependent on its structure and composition. The structure and composition of the major industrial clays, i.e. kaolins, smectites, and palygorskite–sepiolite, are very different even though each is comprised of octahedral and tetrahedral sheets as their basic building blocks. However, the arrangement and composition of the octahedral and tetrahedral sheets account for most differences in their physical and chemical properties.

The important physical and chemical characteristics that relate to the applications of the clay materials are shown in Table 4. Other special properties will be described in the sections on specific clay minerals. In all most all industrial applications, the clays and clay minerals are functional and are not just inert components in the system. In most applications, the clays are used because of the particular physical properties that contribute to the end product, i.e. kaolins for paper coating or bentonite in drilling muds. In some cases, the clay is used for its chemical composition, i.e. kaolin for use as a raw material to make fiberglass or clays and shales in the mix to make cement. The physical and chemical properties

Table 4. Important physical and chemical characteristics of clay materials

Grit percentage (+44 μm)
Particle size, shape and distribution
Mineralogy
Surface area, charge, and chemistry
pH
Ion exchange capacity and identification
Brightness and color
Sorption capacity
Rheology
Ceramic properties
Dispersability

of kaolins, smectites, palygorskite–sepiolite, clays and shales containing illite and chlorite, and refractory and fireclays are discussed in the following sections.

6.1. Kaolins

Kaolin is both a rock term and a mineral term. As a rock term, kaolin means that the rock is comprised predominantly of kaolinite and/or one of the other kaolin minerals. As a mineral term, it is the group name for the minerals kaolinite, dickite, nacrite, and halloysite. Kaolinite is by far the most common kaolin mineral. Dickite, nacrite, and halloysite are relatively rare in comparison. These latter three minerals are commonly formed by hydrothermal alteration, although there are examples of their occurrence in sedimentary and residual deposits in association with kaolinite (Johnson et al., 2000). The term China Clay has been and is used synonymously for kaolin, particularly in Great Britain.

As will be discussed in Chapter 5, for most industrial applications, kaolins must be beneficiated by either dry processing or wet processing in order to reduce or remove impurities and to enhance certain physical properties such as brightness, whiteness, opacity, particle size, shape and distribution, and viscosity. Common impurities in kaolins are quartz, mica, illite, smectite, feldspar, goethite, hematite, pyrite, anatase, rutile, ilmenite, and trace quantities of tourmaline, zircon, kyanite, and a few other heavy minerals. A large percentage of these minerals can be removed by wet processing, which will be discussed in detail in Chapter 4.

Kaolinite has a structure that is comprised of one tetrahedral silica sheet and one octahedral alumina sheet, which are joined by sharing a common layer of oxygens and hydroxyls (Fig. 3). This structure is classed as a 1:1 layer clay. Both the silica tetrahedral sheet and the alumina octahedral sheet have little, if any, substitutions of other elements. Therefore, the charge on the kaolinite layer is minimal, which accounts for several of the physical characteristics shown in Table 5, which is a summary of the properties of kaolinite that relate to its applications.

As shown in the above list, kaolinite is a 1:1 layer clay. In all most all industrial applications, the brightness and color are very important. The brightness and whiteness (color) are two different properties. Brightness is a measure of percentage reflectivity at 457 nm compared to smoked magnesium oxide, which is assigned 100% brightness. (The standard tests for brightness, whiteness, grit percentage, etc. are delineated in Appendix A.) Whiteness or color is measured over the spectrum that is visible to the eye, which is essentially from 400 to 700 nm.The preferred whiteness is blue-white rather than cream-white. However, most kaolins are cream-white and this is referred to as the yellowness factor or *b*-value.

As mentioned before, there is very little substitution of other elements for the aluminum and silicon in the structure and this accounts for many of the properties that are discussed in this section. Ferric iron which has an ionic radius of 0.67 Å could and does have limited substitution for aluminum, which has an ionic radius of 0.57 Å (Newman, 1987). Some aluminum may substitute for silicon in the tetrahedral sheet, but again this substitution is very limited. Some believe that there is limited substitution of silicon by titanium, which has an ionic radius of 0.64 Å (Jepson and Rowse, 1975). Weaver (1976), however, presented evidence that titanium in kaolinite is present as discrete surface-sorbed forms. Thus, the charge on the lattice is minimal, particularly on the basal (001) surfaces. The major charge on the kaolinite particle is caused by broken bonds along the edges (Grim, 1962).

Because of the limited substitution in the kaolinite lattice, the base exchange capacity and the sorptivity are low in comparison to smectites

Table 5. Physical characteristics of kaolinite

1:1 Layer clay
White or near-white in color
Very limited substitutions in the structure
Minimal charge on the layer
Very low cation exchange capacity
Pseudo-hexagonal plates and books (Fig. 9)
Relatively low surface area
Low absorption capacity
Good rheology
Refractory
Plastic

and palygorskite–sepiolite. Typically, the base exchange capacity of kaolinite is in the range of 1–5 meq/100 g. Kaolinite exhibits low absorption and adsorption properties, which are directly related to the low surface charge on the particle.

The morphology of the kaolinite particles, as shown previously in Fig. 6, shows well-defined pseudo-hexagonal plates and in some deposits, relatively thick books or stacks and some long-vermicular-shaped crystals. Some relatively pure kaolin deposits, such as those that occur in Georgia in the United States and in the states of Amapa and Para in Brazil, exhibit good rheology or flow characteristics at high solids concentrations. There are several factors that affect the rheology (generally measured at 70% solids; see Appendix A). Some of these factors are listed in Table 6.

The reasons that some kaolins have good rheology are that there is little or no charge deficiencies in the structure, they have a relatively low surface area (8–15 m^2/g), exhibit good crystalline morphology, and are fine, but have a relatively broad particle size distribution (Fig. 16). Fig. 17 shows the relationship between particle packing and viscosity.

Another beneficial property of kaolinite is that it is soft and thus is non-abrasive. The hardness on the Mohs' scale is about 1.5. This property is very important in many industrial applications because the kaolin is softer than almost all the materials with which it comes into contact, and therefore, the wear and tear on equipment and machinery is low.

Relatively pure kaolins are refractory and melt or fuse at a temperature of about 1850 °C. In most instances, except for flint clays, kaolins are plastic, fire with a high modulus of rupture and to a white or near-white color. These properties make kaolin a very important ceramic raw material, as will be discussed in detail in Chapter 5.

Other physical and chemical properties that may be important are that kaolin is chemically inert over a relatively wide pH range (4–9), has low conductivity of both heat and electricity, is hydrophilic and disperses

Table 6. Factors that affect viscosity

Particle size
Particle shape
Particle size distribution
Presence of impurities such as smectite and illite
Presence of halloysite
Presence of fine mica or illite
Soluble salt content
Crystal perfection

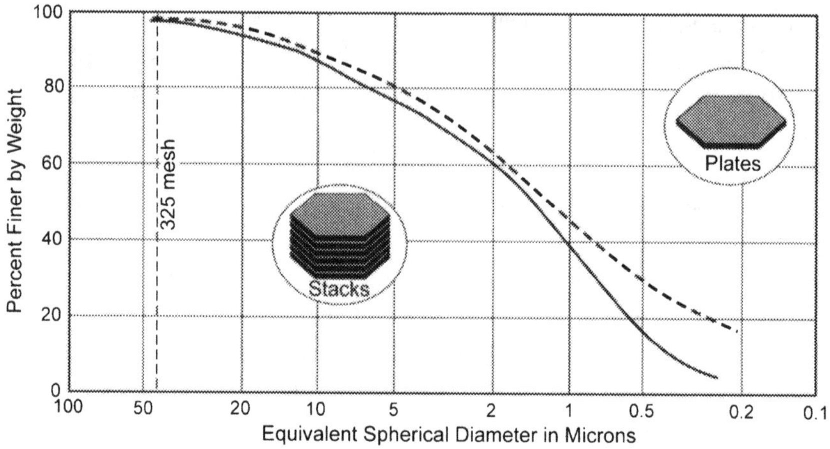

Fig. 16. Typical particle size distribution of soft kaolins: Georgia (dashed line) and Brazil (solid line).

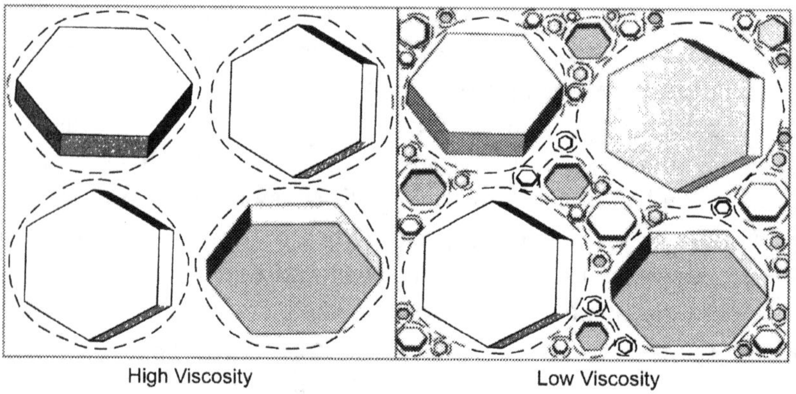

Fig. 17. Diagrammatic representation of the relationship between particle packing and viscosity.

readily in water, and can be thermally treated or calcined to produce products that are excellent fillers and extenders, which will be discussed in Chapters 4 and 5.

6.2. *Smectites*

Smectite is the group name for several hydrated sodium, calcium, magnesium, iron, and lithium aluminum silicates. The individual mineral names in the group are sodium montmorillonite, calcium montmorillonite, saponite (Mg), nontronite (Fe), and hectorite (Li). The rock term bentonite is commonly used for these minerals and was defined by Ross and Shannon (1926) as a clay material altered from a glassy igneous material, usually volcanic ash. Grim and Guven (1978) used the term bentonite for any clay which was dominantly comprised of a smectite mineral without regard to its origin. Those bentonites that are used industrially are predominantly comprised of either sodium montmorillonite, calcium montmorillonite, or, to a much lesser extent, hectorite.

Smectites are three-layer minerals (Fig. 8) in contrast with kaolinite which is a two-layer mineral. This three-layer clay has two silica tetrahedral sheets joined to a central octahedral sheet. There can be considerable substitution in the octahedral sheet of Fe^{3+}, Fe^{2+}, and Mg^{2+} for Al^{3+}, which creates a charge deficiency in the layer. Also, there can be some substitution of silicon by aluminum in the tetrahedral sheets, which again creates a charge imbalance. Grim (1962) pointed out that many analyses have shown that this charge imbalance is about −0.66 per unit cell. This net positive charge deficiency is balanced by exchangeable cations adsorbed between the unit layers and on the edges. Thus, if the exchangeable cation is sodium, the specific mineral is sodium montmorillonite and if it is calcium, it is a calcium montmorillonite. According to Grim (1962), substitution within the lattice causes about 80% of the total cation exchange capacity and broken bonds around the edges of the particles, about 20%. Sodium and calcium ions are hydrated, and when in the interlayer position, sodium montmorillonites have one associated molecular water layer and calcium montmorillonites generally have two associated molecular water layers. This results in sodium montmorillonites having a *c*-axis spacing of about 12.3 Å and calcium montmorillonites having a *c*-axis spacing of about 15 Å (Grim and Guven, 1978).

Table 7 is a summary of characteristics of smectites which relate to their applications.

As mentioned above, the three smectite varieties that are most used industrially are sodium montmorillonite, calcium montmorillonite, and hectorite. Sodium montmorillonite and hectorite have high base exchange capacities, generally ranging between 80 and 130 meq/100 g. Calcium montmorillonite, on the other hand, has a base exchange capacity that normally ranges between 40 and 70 meq/100 g. The high charge on the lattice gives both sodium montmorillonite and hectorite the capacity to exchange the interlayer water and associated cations with more polar organic molecules such as ethylene glycol, quaternary amine, and polyalcohols. This is an important property, which can be translated into very useful products called organoclays. Sodium montmorillonite is comprised of very small thin flakes (Fig. 9). This has been described by Keller (1982) as cornflake texture. This results in the sodium montmorillonites

Table 7. Characteristics of smectite

2:1 Layer clay
Variable color, usually tan or greenish-gray
Considerable lattice substitutions
High layer charge
Medium to high cation exchange capacity
Very fine particle size
High surface area
High sorptive capacity
High viscosity
Thixotropic
Very low permeability
Medium to high swelling capacity
High green and dry compression strength
High plasticity

having a very high surface area of about 150–200 m^2/g. The high surface area and high layer charge give sodium montmorillonite a high sorptivity and a very high viscosity at low solids concentration (5%), in contrast to kaolinite, which has a low viscosity at 70% solids.

Because of the high layer charge and base exchange capacity, sodium montmorillonites have a high swelling capacity of the order of 10–15 times when placed in water. This swelling capacity results in many beneficial uses, which will be described in Chapter 6. A unique property of sodium montmorillonite and hectorite is that of thixotropy. Thixotropy is the ability to form a gel upon standing and to become fluid when stirred or agitated. This property makes sodium montmorillonite and hectorite excellent suspending agents. Sodium montmorillonite is the premier drilling mud and hectorite is used in pharmaceutical and medicinal suspensions and in some high quality paint.

The very fine particle size, swelling capacity, and flake shape give sodium montmorillonite the ability to form almost impermeable membranes to the movement of water. This makes it a very good sealant for use to line irrigation ditches and landfills and to form an impermeable seal on permeable formations when drilling oil and gas wells, to prevent fluid loss.

Calcium montmorillonite is generally larger in particle size, has a lower surface area (50–80 m^2/g), a lower base exchange capacity, a lower swelling index (2–3), and a lower viscosity than sodium montmorillonite. These properties can be increased by exchanging the calcium with sodium, but rarely do the properties equal those of a natural sodium montmorillonite.

Both calcium and sodium montmorillonites have good properties needed for bonding sands in foundry molds. Those properties are green compression strength, dry compression strength, hot strength, flowability, permeability, and durability (Grim and Guven, 1978). Calcium montmorillonite has a higher green strength, lower dry strength, lower hot strength, and better flowability than sodium montmorillonite. Thus the physical and chemical properties of these smectites are largely controlled by particle size, particle shape, lattice substitutions, and surface area. The color of the bentonites are variable, ranging from tan to blue-gray, olive, brown, and, rarely, white. White bentonites are very rare in occurrence and are more valuable in many applications that desire a white color.

6.3. Palygorskite–Sepiolite

As mentioned earlier, palygorskite and attapulgite are names for the same hydrated aluminum silicate mineral. Sepiolite is structurally similar to palygorskite except that it has a slightly larger unit cell. Both minerals consist of double silica tetrahedral chains linked by octahedral oxygen and hydroxyl groups containing aluminum and magnesium ions in a chain-like structure (Fig. 14). The morphology of both minerals is an elongate chain or lath-like particle (Fig. 15).

The term fuller's earth is a catchall for clays or other fine-grained, earthy materials suitable for bleaching and sorptive uses. Fuller's earth has no compositional or mineralogical connotation. It was first applied to earthy materials used in cleansing and fulling wool, which removed lanolin and dirt, thus the name fuller's earth (Robertson, 1986). The term is therefore quite often used for the palygorskite (attapulgite), sepiolite, and smectite clays that have natural bleaching and/or sorptive qualities.

The physical and chemical characteristics of palygorskite and sepiolite are summarized in Table 8.

As shown in Fig. 14, both palygorskite and sepiolite have 2:1 inverted structures, i.e. the apices of the silica tetrahedrons are regularly inverted along the *a*-axis. This results in parallel channels throughout the particles which gives these minerals a high internal surface area. The base exchange capacity is intermediate between kaolinite and sodium and calcium montmorillonites, usually of the order of 30–40 meq/100 g.

The fine particle size, high surface area (190 m^2/g), and medium exchange capacity give both palygorskite and sepiolite a high capacity to absorb and adsorb various liquids, which make them very useful in many industrial applications. Another desirable characteristic is that the elongate thin particles cause high viscosity when added to any liquid. It is a physical and not a chemical viscosity, so is very stable as a viscosifier and

Table 8. Characteristics of palygorskite–sepiolite

2:1 Layer clay
Light tan, brown, cream, or bluish green color
Elongate, very thin particles
Some lattice substitutions
Moderate layer charge
Medium base exchange capacity
High surface area
High sorptive capacity
High viscosity

suspending medium in many applications where sodium montmorillonite would flocculate when the salt or electrolyte concentration is high. Many applications related to sorption and viscosity are discussed in Chapter 7.

6.4. Common Clays

Common clays can be seat earths (underclays), shales, lacustrine clays, soils, and other clay-rich materials (Murray, 1994). Usually, the clay mineral composition of these materials is mixed. For example, shales commonly contain illite (Fig. 10), chlorite (Fig. 12), and mixed-layer illite–smectite (Fig. 18) or illite–chlorite. Mixed-layered or interstratified clay minerals usually contain two components such as illite and smectite. Most commonly, the layers are randomly ordered, but can be regularly ordered. This regularly ordered illite–smectite is called rectorite (Moore and Reynolds, 1997). The physical and chemical properties are very diverse, so these common clays are utilized for specific end uses. The physical properties that are normally important relate to their use in the manufacture of structural clay products such as bricks and tiles. These properties are plasticity, green strength, dry strength, dry and fired shrinkage, fired color, fired strength, vitrification range, and fired density. Many shales and seat earths (underclays) are suitable for making structural clay products.

Certain low grade refractories can be made from some underclays or fireclays. These clays are generally mixtures of predominantly kaolinite, along with minor quantities of illite and/or chlorite. For this use, in addition to those physical properties needed for a common brick clay, the pyrometric cone equivalent (PCE) is important, along with the ability to withstand moderately high temperatures without melting or fusing. Refractory bricks are classified as low, medium, high, and super duty. The PCE values and temperatures for these refractory grades are shown in Chapter 5.

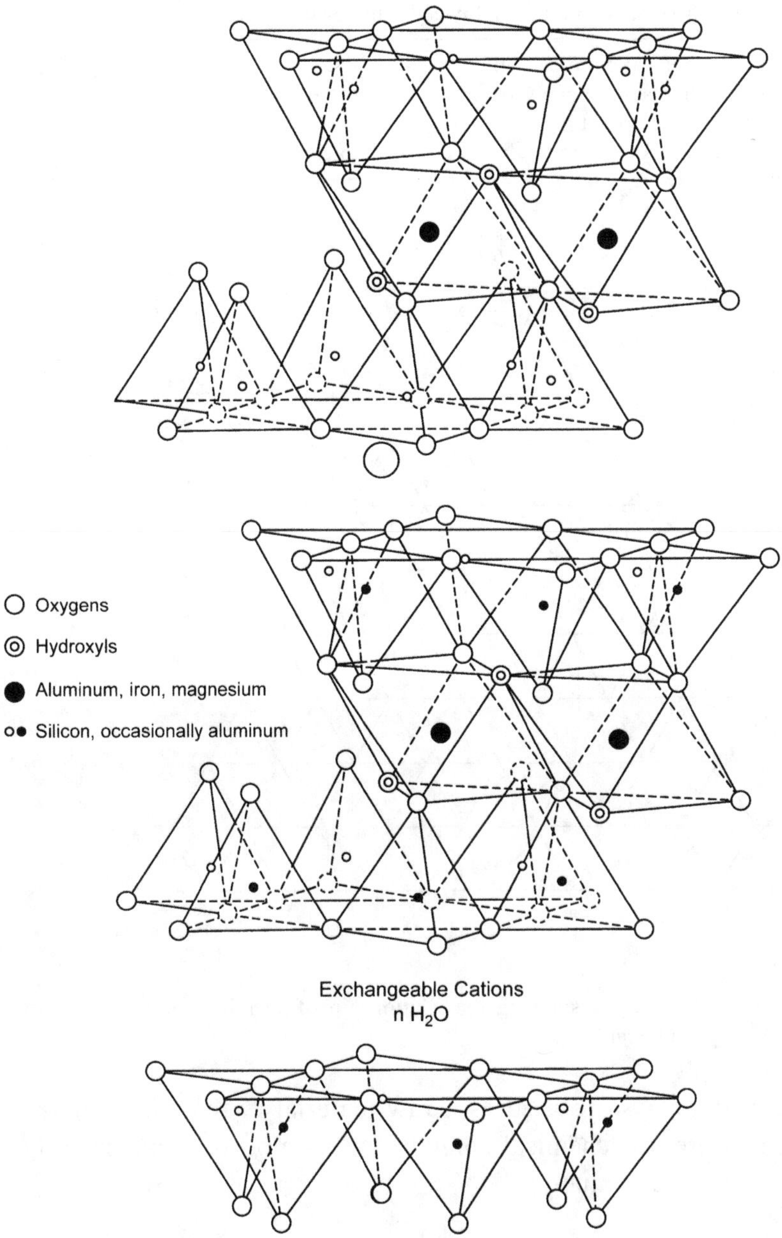

Fig. 18. Diagrammatic sketch of mixed-layer illite and smectite.

Some common clays or shales are used to make lightweight aggregate (Murray and Smith, 1958; Mason, 1994). The important physical properties necessary for this purpose are shown in Table 9. Riley (1951) pointed out the chemical properties necessary to produce a bloating clay (Fig. 19).

Table 9. Properties necessary to make lightweight aggregate

Chemical and mineralogical composition
Temperature at which gases are evolved
Vitrification range and temperature
Fired specific gravity (this use will be discussed in Chapter 8)

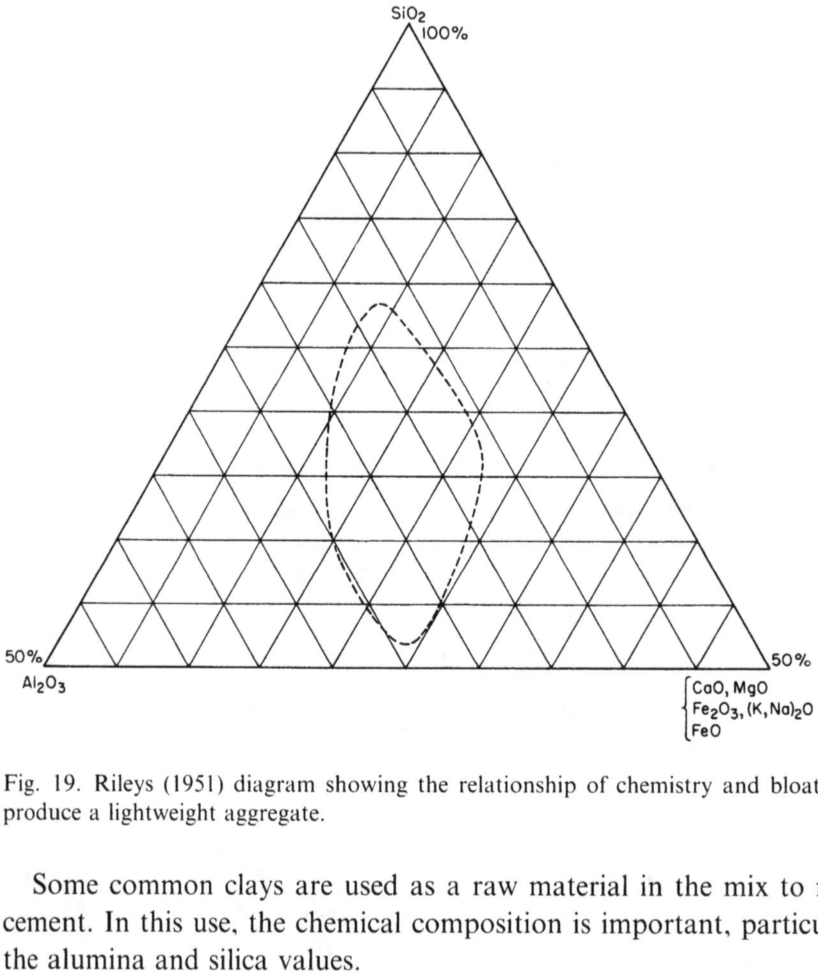

Fig. 19. Rileys (1951) diagram showing the relationship of chemistry and bloating to produce a lightweight aggregate.

Some common clays are used as a raw material in the mix to make cement. In this use, the chemical composition is important, particularly the alumina and silica values.

REFERENCES

Bailey, S.W. (1980) Structures of layer silicates. Chapter 1 in *Crystal Structures of Clay Minerals and their X-Ray Identification*. Brindley, G.W. and Brown, G., eds. Mineralogical Society Monograph No. 5, London, pp. 1–123.

Bailey, S.W. (1988) Chlorites: structures and crystal chemistry. Chapter 10 in *Hydrous Phyllosilicates*. Bailey, S.W., ed. Reviews in Mineralogy Vol. 19, pp. 347–403.

Bailey, S.W. (1993) Review of the structural relationships of the kaolin minerals. Chapter in *Kaolin Genesis and Utilization*. Murray, H., Bundy, W., and Harvey, C., eds. Special Publication No. 1. The Clay Minerals Society, Boulder, CO, pp. 25–42.

Bates, T.F., Hildebrand, F.A., and Swineford, A. (1950) Morphology and structure of endellite and halloysite. *Am. Mineral.*, 35, 463–484.

Grim, R.E. (1962) *Applied Clay Mineralogy*. McGraw-Hill, New York, 422pp.

Grim, R.E. (1968) *Clay Mineralogy*, 2nd Edition. McGraw-Hill, New York, 596pp.

Grim, R.E., Bray, R.H., and Bradley, W.F. (1937) The mica and argillaceous sediments. *Am. Mineral.*, 22, 813–829.

Grim, R.G. and Guven, N. (1978) *Bentonites—Geology, Mineralogy, Properties and Uses*. Elsevier, Amsterdam, 256pp.

Guven, N. (1988) Smectites. Chapter 13 in *Hydrous Phyllosilicates*. Bailey, S.W., ed. Reviews in Mineralogy Vol. 19, pp. 497–559.

Jepson, W.B. and Rowse, J.B. (1975) The composition of kaolinite—an electron microscope microprobe study. *Clay. Clay Miner.*, 23, 310–317.

Johnson, C.T., et al. (2000) Distribution of dickite and nacrite stacking sequences in kaolin (Abstract). Clay Minerals Society, *37th Annual Meeting*, Loyola University, Chicago, IL, p. 69.

Jones, B.F. and Galan, E. (1988) Sepiolite and palygorskite. Chapter 16 in *Hydrous Phyllosilicates*. Bailey, S.W., ed. Reviews in Mineralogy Vol. 19, pp. 631–674.

Keller, W.D. (1982) Applications of scanning electron microscopy to clays and other fine-grained minerals. *Proceedings Process Mineralogy II*. Hagni, R.D., ed. Metallurgical Society, AIME, pp. 245–261.

Mason, B.H. (1994) Lightweight aggregate. Chapter in *Industrial Minerals and Rocks*, 6th Edition. Carr, D.D., ed. Society for Mining, Metallurgy and Exploration, Littleton, CO, pp. 343–350.

Moore, D.M. and Reynolds, R.C. Jr. (1997) *X-Ray Diffraction and the Identification and Analysis of Clay Minerals*, 2nd Edition. Oxford University Press, Oxford and New York, 378pp.

Murray, H.H. (1994) Common clay. Chapter in *Industrial Minerals and Rocks*, 6th Edition. Carr, D.D., ed. Society for Mining, Metallurgy and Exploration, Littleton, CO, pp. 247–248.

Murray, H.H. and Smith, J.M. (1958) *Lightweight Aggregate Potentialities of Some Indiana Shales*. Indiana Geological Survey Report of Progress No. 12, 42pp.

Newman, A.C.D., ed. (1987) *Chemistry of Clays and Clay Minerals*. Mineralogical Society Monograph No. 6, London, 480pp.

Riley, C.M. (1951) Relation of chemical properties to the bloating of clays. *J. Am. Ceram. Soc.*, 34, 121–128.

Robertson, R.H.S. (1986) *Fuller's Earth—A History*. Volturna Press, Hythe, Kent, UK, 421pp.

Ross, C.S. and Shannon, E.V. (1926) Minerals of bentonite and related clays and their physical properties. *J. Am. Ceram. Soc.*, 9, 77–96.

Weaver, C.E. (1976) The nature of T_iO_2 in kaolinite. *Clay. Clay Miner.*, 24, 215–218.

Chapter 3

GEOLOGY AND LOCATION OF MAJOR INDUSTRIAL CLAY DEPOSITS

There are many clay deposits around the world that are mined and processed for industrial uses. However, in this chapter, only the major world-class deposits of kaolins (including halloysite and ball clay), bentonites, and palygorskite–sepiolite that are marketed regionally or worldwide are described.

1. KAOLINS

1.1. United States

The sedimentary kaolins in Georgia and South Carolina have been utilized since the middle 1700s. The total tonnage mined to date from this area is over 500,000,000 tons. This region is one of two of the most extensive areas of secondary kaolin deposits in the world. These kaolins occur as Late Cretaceous and Early Tertiary age lenses and beds which were derived from weathered granites, gneisses, and phyllites on the Piedmont Plateau (Fig. 20). In Late Cretaceous time, the residual weathering products of granites and gneisses were eroded and transported to the coastline located along what is known as the fall line. The fall line is the boundary between the crystalline rocks of the Piedmont Plateau and the Coastal Plain sediments to the southeast (Fig. 20). These detrital weathering products were deposited in lagoons, estuaries, oxbows, lakes, and ponds in a broad deltaic system covering an area about 30–50 km wide and 150 km long extending from Macon, Georgia to Aiken, South Carolina (Kogel et al., 2002). The kaolin beds range from 2 to 12 m thick. The Tertiary kaolins are generally larger in areal extent and thicker than the Cretaceous kaolins (Murray and Keller, 1993).

The Cretaceous kaolins are relatively coarse with a particle size ranging from 55% to 75% $<2\,\mu m$. They are often referred to as soft kaolins. The Tertiary kaolins are much finer with a particle size finer than about 85% $<2\,\mu m$ or finer and are called hard kaolins. In the book by Kogel et al., (2002), the typical properties of the soft and hard kaolins are

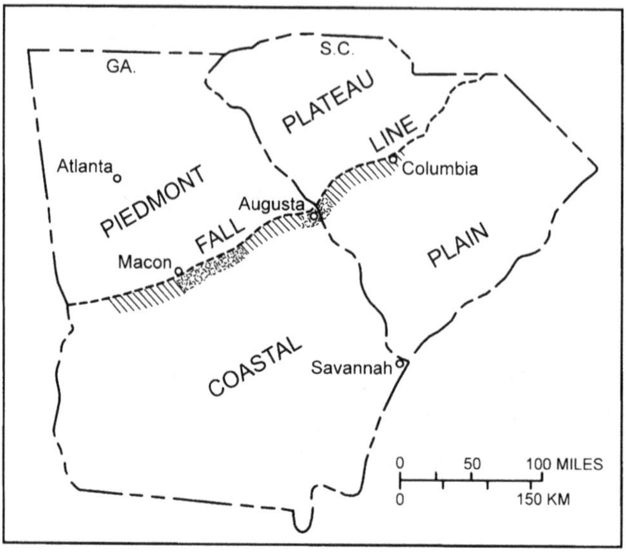

Fig. 20. Location map of kaolins in Georgia and South Carolina.

summarized. Fig. 21a, b shows typical electron micrographs of a soft Cretaceous kaolin and a hard Tertiary kaolin. The soft kaolins contain a large quantity of coarse stacks interspersed in a matrix of finer platy particles. The hard kaolins contain thin platy particles with no large books or stacks (Murray and Keller, 1993). The Hinckley index of crystallinity (Hinckley, 1963) is high for the soft kaolins and low for the hard kaolins. Dombrowski (1993) reviewed the theories for the origin of the Cretaceous soft kaolins and the Tertiary hard kaolins. His research showed that the Cretaceous kaolins were derived from granites and gneisses and the Tertiary kaolins were derived from phyllite source material. These results were based on trace element geochemistry.

Kaolinite is the dominant mineral present in these Georgia and South Carolina kaolins, generally comprising more than 90% of the mineral content (Murray, 1976). Other accessory minerals that are commonly present include quartz, muscovite, biotite, partially altered feldspar, and smectite along with minor amounts of halloysite, magnetite, ilmenite, rutile, anatase, zircon, tourmaline, and a few other heavy minerals previously enumerated in Chapter 1. The current annual production is about 7,400,000 tons as estimated by the US Geological Survey (Virta, 2004).

1.2. England

The largest and highest quality primary kaolin deposits in the world are located in Cornwall and Devon in southwestern England (Fig. 22). These

(a)

(b)

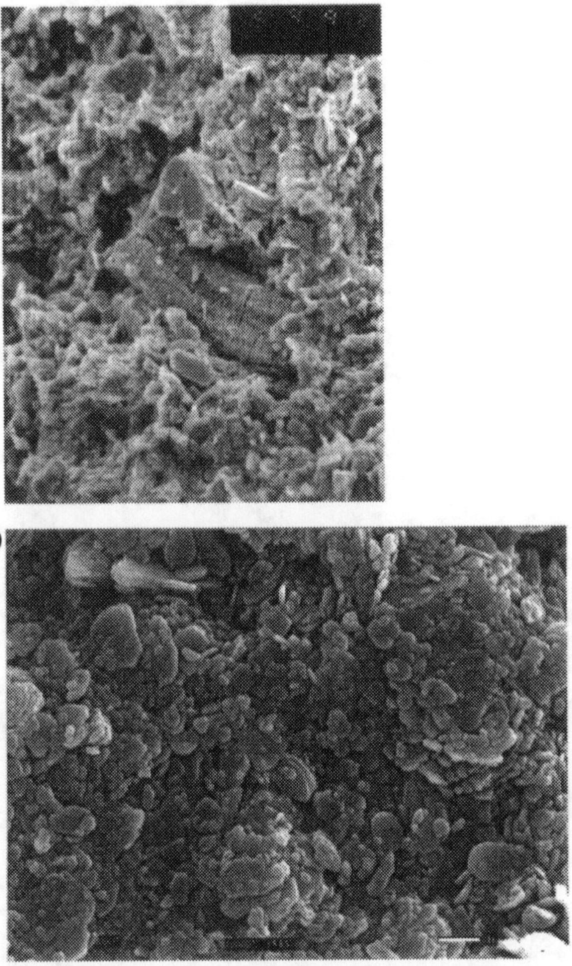

Fig. 21. Scanning electron micrograph of (a) a soft Cretaceous Georgia kaolin and (b) a hard Tertiary Georgia kaolin.

kaolins were discovered in 1746 (Wilson, 2002). Wilson estimates that the total production since their discovery has been about 170,000,000 tons. The kaolinite is derived from late stage magmatic or hydrothermal decomposition of feldspar in the granite (Bristow, 1993). The kaolinite content in the altered granite ranges between 10% and 20%. Extensive drilling has shown that the typical kaolin body is funnel- or trough-like in form (Fig. 23). Dewu and Durrance (1993) have shown that uranium is the main source of radiogenic heat that has resulted in a hydrothermal circulation system which is an on-going process that is still actively forming kaolinite within the granites. Two distinct types of kaolinite occur in the granites: a fine platy type and a vermiform type (Wilson,

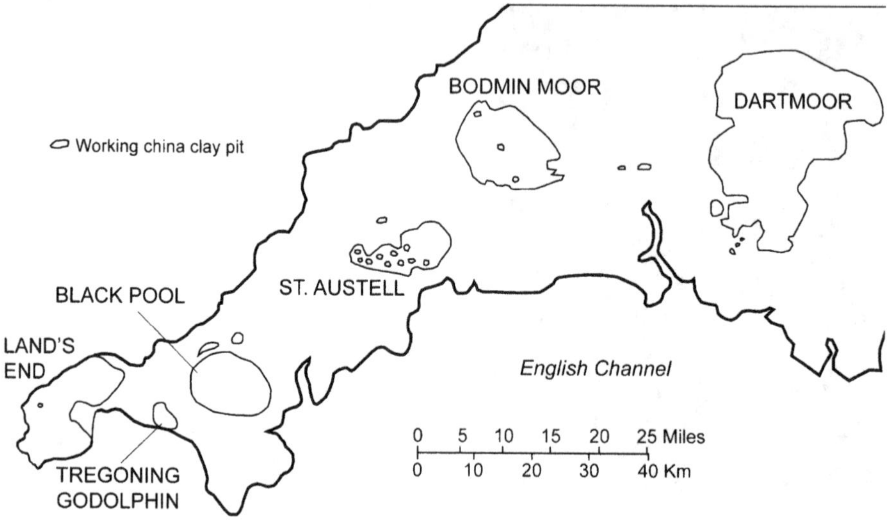

Fig. 22. Kaolin deposits in southwestern England.

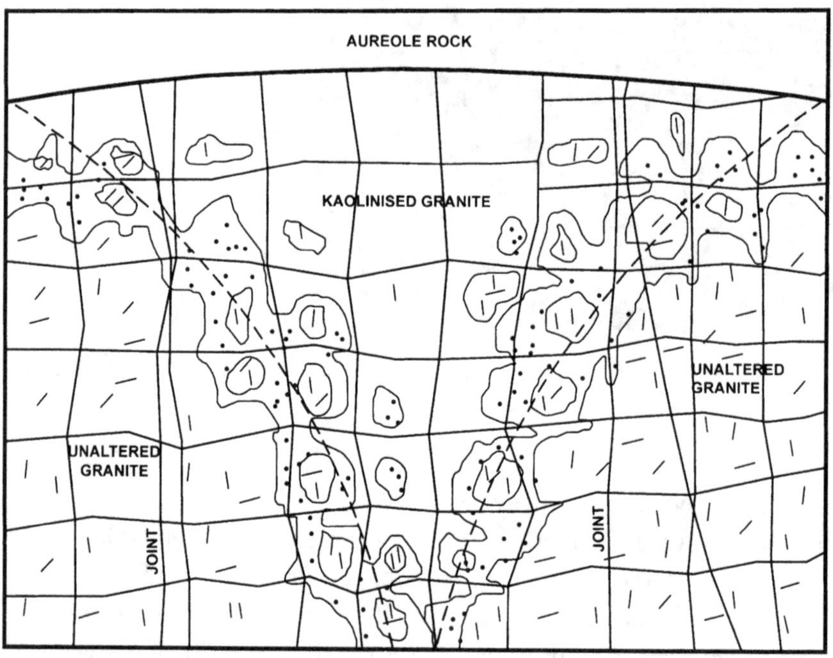

Fig. 23. Typical shape of the hydrothermal English kaolin deposits.

2002). The fine platy kaolinite is associated with feldspar and the vermiform kaolinite is derived from muscovite. Because of the low iron content of the parent granite, the recovered kaolinite is very white. The current annual production is about 2,000,000 tons (Wilson, 2002).

1.3. Brazil

The Amazon region in northern Brazil has two areas where sedimentary kaolins are mined and processed primarily for use by the paper industry (Fig. 24). The first area on the Jari River, a tributary of the Amazon, was discovered in 1970. The kaolin in this large deposit is Pliocene in age and is called the Belterra clay (Murray, 1981). The extent of this deposit is 12 km in length and 5 km in width and is up to 40 m thick. The top 8 m of the kaolin is laterized and contains considerable gibbsite. The kaolin below the laterite is very fine, normally at least 90% <2 μm, so is a hard kaolin in which the kaolinite particles are small plates with no books or stacks (Fig. 25). The kaolin occurs near the top of the Barreiras series (Fig. 26), and is located on a plateau on the east side of the Jari River. This plateau is dissected by several streams and is overgrown with a dense tropical forest. The elevation of the plateau is about 150 m above the Jari River. The source of the Belterra kaolin was from the crystalline rocks on the Guyana Shield about 15 km north (Murray and Partridge, 1982). The current annual production of this Jari kaolin is about 850,000 tons (Franca, 2002).

The second area where sedimentary kaolins are being mined and processed is along the Capim River south of Belem in the state of Para (Fig. 24). This Capim basin is located between the 2° and 3° parallels and the 47° and 48° meridians just south of the equator. The kaolin layers occur in the Ipixuna Formation of Tertiary age (Fig. 27). The kaolin

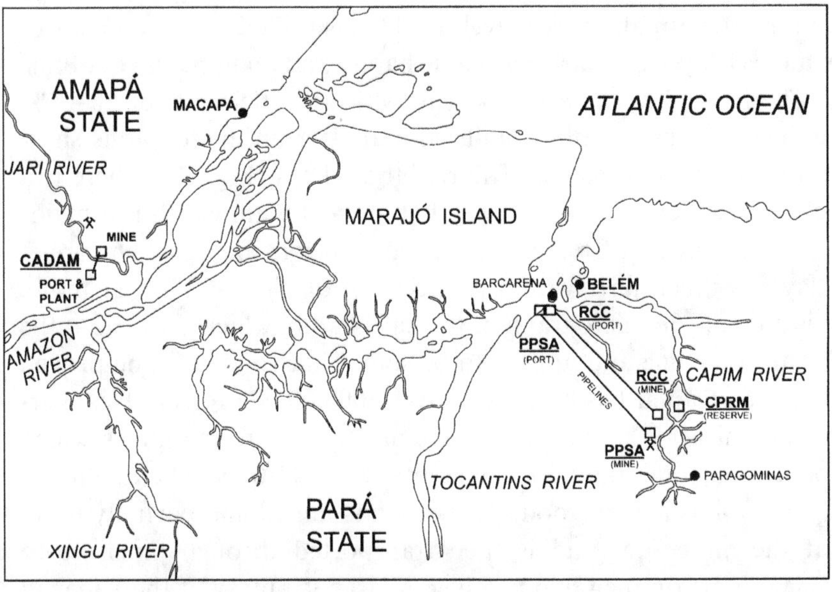

Fig. 24. Location of kaolin mines and plants in northern Brazil.

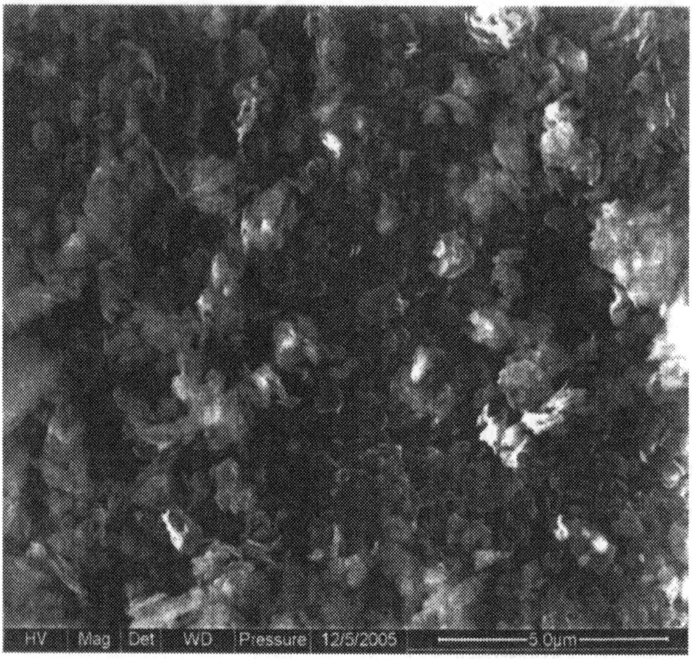

Fig. 25. Scanning electron micrograph of the Jari kaolin.

deposits are found in low, slightly elevated plateaus with dissected edges which have very little regional expression and are somewhat difficult to distinguish (Alves, Personal communication). These low lying plateaus range from 2 to 3 km in length and width and show a difference in elevation of 30–60 m above sea level. In the basin there are two layers of kaolin, a lower layer of relatively coarse kaolin and an upper layer of fine kaolin. The lower layer is a soft kaolin with a particle size of 50–65% <2 μm. This kaolin is well crystallized and electron micrographs show many large stacks in a matrix of fairly large plates (Fig. 28). There is a relatively low percentage of particles that are <0.5 μm which is a major difference from the soft Cretaceous age kaolins from Georgia. The upper kaolin layer is much finer with a particle size of 85% <2 μm or more. Fig. 29 is an electron micrograph of the hard kaolin which shows that the particles are very fine and appear to be somewhat rounded. The proven reserves in the Capim basin are well over 500,000,000 tons (Alves, Personal communication). Currently, there are two operating mines which are producing a total annual tonnage of 1,600,000 tons. The quality of the Capim kaolins is very good. The kaolin is mined and partially processed at the mine sites and is then transported through pipelines to Barcarena, a port on the Guama River, a large tributary of the Amazon (Fig. 24), where the kaolin is prepared for slurry or dry shipment.

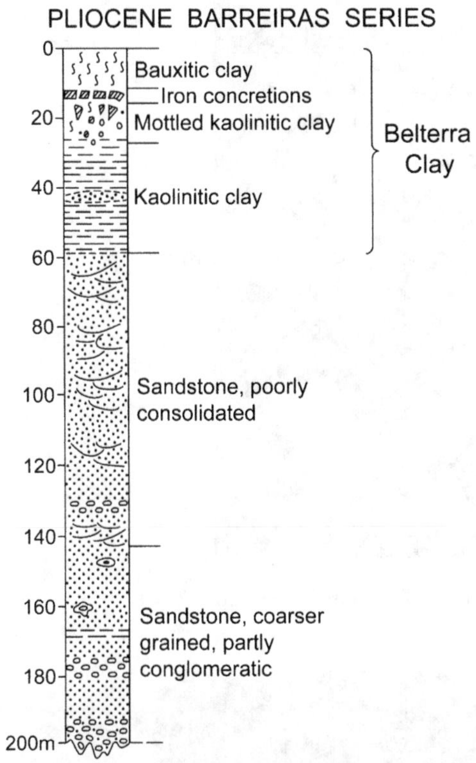

Fig. 26. Stratigraphic column of Pliocene Barreiras series.

	AGE	FORMATION	ROCK UNITS	FACIES UNITS
Quaternary	Holocene	Alluvial Terrain		Sand and gravel units
Quaternary	Pleistocene	Barreiras Formation	1, 2, 3	1 - Gravel quartz pebbles 2 - Yellow sandstone 3 - Interbedded argillaceous-sand layers
Tertiary	Pliocene Miocene	Ipixuna Formation	1, 2	1 - Kaolinic crossbedded sandstone with minor interbedded siltstone 2 - Thin interbedded kaolin layers related with kaolinic sandstone (ore zone)
	Cretaceous	Itapecuru Formation	1, 2	1 - Thin bedded pink cross-bedded sandstone, siltstone, and gravel 2 - Arcosean cross-bedded sandstone with siltstone related
	Eo-Paleozoic	Guama Formation		White fine sandstone, bad sorted. Parallel beds laminations and bioturbated related

Fig. 27. Stratigraphic column showing the Ipixuna Formation.

1.4. Czech Republic

The kaolin deposits in the Czech Republic are mainly residual deposits which were formed by weathering of granites in the Karlovy Vary region (Fig. 30). Also, a carboniferous arkose in the Podborany area has been

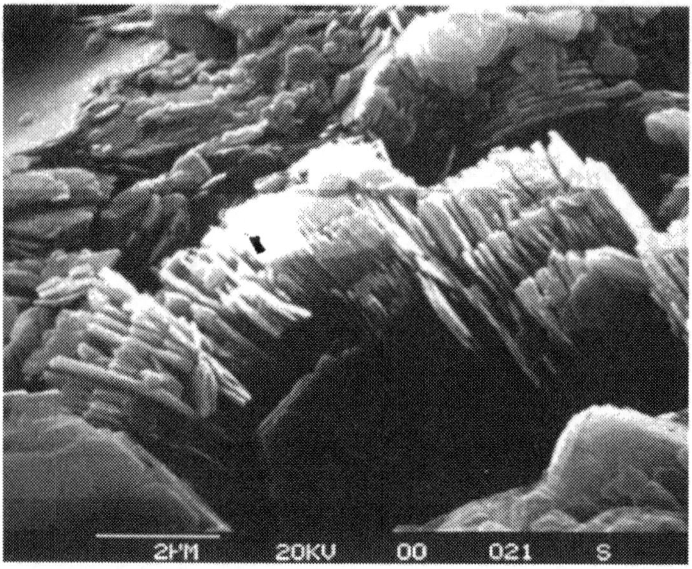

Fig. 28. Scanning electron micrograph of the soft Capim basin kaolin.

Fig. 29. Scanning electron micrograph of the hard Capim basin kaolin.

kaolinized by residual weathering (Kuzvart, 1984). The largest deposit is located between Podborany and Krasny Dvur and contains over 100 million tons of reserves (Aron, 2000). The average thickness of this deposit is 30 m. In the Karlovy Vary area, there is a large pluton called the western Krusne Hory Pluton, which is comprised of lithium micaceous granites. These granites are highly jointed and faulted so that surface residual

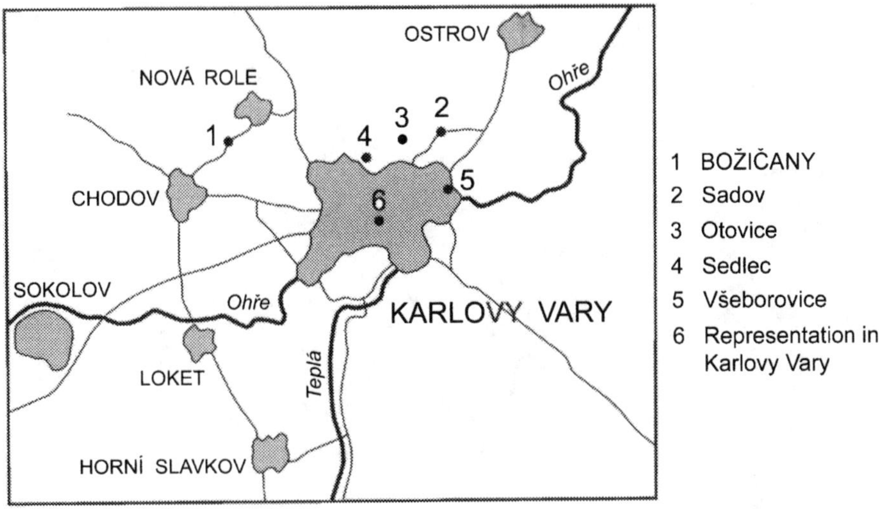

Fig. 30. Location of kaolin deposits in the Czech Republic near Karlovy Vary.

kaolinization can be very deep (Krelina et al., 1982). The principal peak of climatic kaolinization took place in Cretaceous and Paleocene times (Neuzil and Konta, 1965). Zettlitz kaolin has been supplied to the porcelain and ceramics industries in Europe since 1805. The original Zettlitz deposit near the village of Sedlec was abandoned in 1962. However, other primary deposits in the area are similar in quality and now are used to supply Zettlitz kaolin primarily to ceramic customers. The annual production of washed kaolin from the Czech Republic is estimated to be 750,000 tons (Wilson, 2002).

1.5. Indonesia

Kaolin is mined on the island of Belitung and Bangka located in the Java Sea north of Jakarta (Fig. 31). The majority of the mining is on Belitung. The kaolins there are residual and altered from porphyritic granite (Murray et al., 1978). The mineralogical and chemical changes in the altered granite from the surface down to the fresh granite show that the feldspars alter to kaolinite and halloysite; the biotite alters to hydrobiotite and then to vermiculite followed by smectite; and the quartz is unaltered except for solution pitting. The kaolin which is mined is comprised of vermicular kaolinite stacks, and plates along with some halloysite (Fig. 32).

The granite which is altered to kaolinite underlies the northwest quarter of the island (about 625 km^2) and is Triassic in age. Since Cretaceous time, the area has been very stable so that intensive tropical weathering altered the granite to depths averaging about 8 m. This residual kaolin

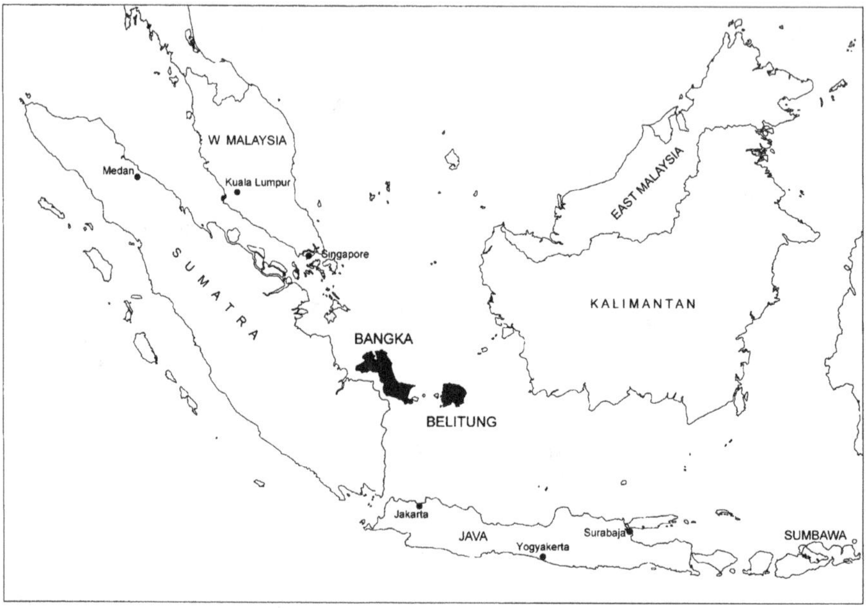

Fig. 31. Location of Bangka and Belitung where kaolin deposits occur.

Fig. 32. Scanning electron micrograph of Belitung kaolin.

has a low iron and titanium content, so has a relatively high brightness. The kaolin is used locally as a ceramic raw material and is shipped to Japan, the Philippines, and Taiwan, where it is used as a filler in paper. Because of the halloysite and smectite content, the viscosity is high so it is not useable as a paper coating clay. Wilson (2002) estimated the current annual production at 500,000 tons.

1.6. China

There are many primary kaolin deposits in China, the largest of which is a hydrothermal deposit at Suzhou. Suzhou is located southwest of Shanghai in Jiangsu Province (Fig. 33). The kaolin is altered from Jurassic age volcanics. The Suzhou deposits are very complex because of the diverse parent rocks, geologic structures, and the hydrothermal and groundwater leaching (Zheng Zhi et al., 1982). The kaolinized zone averages 20 m in thickness and is comprised mainly of kaolinite, halloysite, and quartz, along with minor amounts of smectite, sericite, alunite, and pyrite. The Suzhou kaolin is used as a ceramic raw material and a filler in paper and

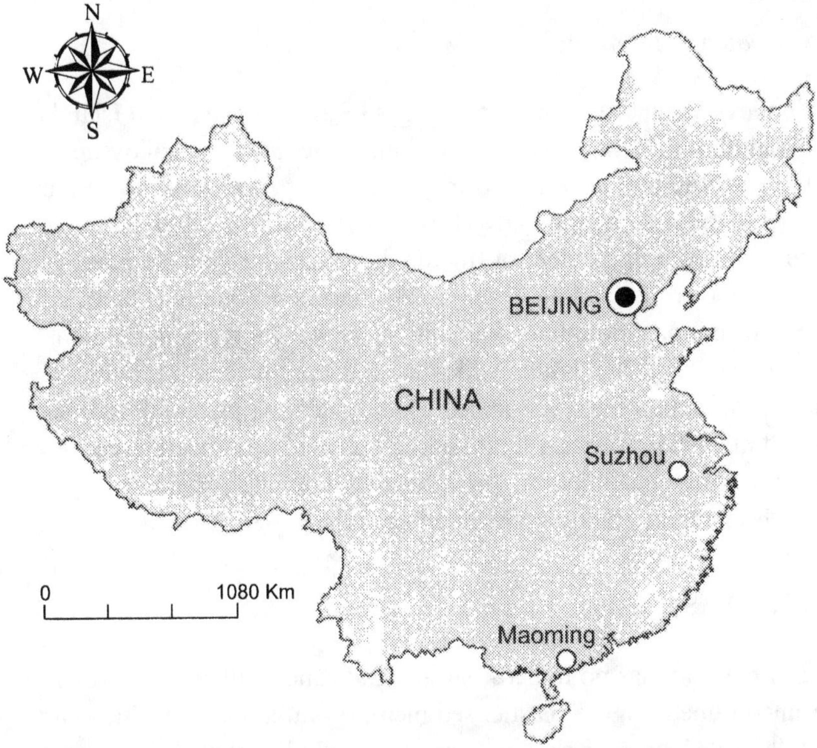

Fig. 33. Location of Suzhou and Maoming kaolin deposits.

rubber and an extender in paint. Other primary kaolins in China are used locally mainly as a ceramic raw material.

The only large sedimentary deposit in China is located near Maoming in western Guandong Province (Fig. 33). This kaolin is actually a kaolinitic sand of Late Tertiary age. The Bureau of Geology and Mineral Resources of Guandong estimated the kaolin reserves in this large kaolinitic sand are over 200 million tons. The kaolinitic sand body ranges from 20 to 45 m in thickness and extends horizontally for about 50 km (Yuan and Murray, 1993). The mineral content of this large Tertiary body is 75–80% quartz and 20–25% kaolinite. The genesis of the Maoming kaolin is attributed to in situ alteration of feldspar in an original arkosic sand. The arkosic rocks were derived from intensive weathering of Precambrian gneisses and Cretaceous acidic igneous rocks which were transported a short distance and deposited in the Maoming sedimentary basin (Lin et al., 1989). The kaolinite recovered from the friable kaolinitic sand has a particle size of 95% finer than 10 μm and 60% finer than 2 μm (Yuan, 1994). The kaolinite particles are platy and have a sharp pseudo-hexagonal morphology. This kaolin is used as a ceramic raw material and as a filler and coating clay for paper. Wilson (2002) estimates the annual production of kaolin in China is about 2,000,000 tons.

1.7. New Zealand Halloysite

The only halloysite mine currently operating is located on the North Island of New Zealand about 240 km north of Auckland (Fig. 34). The halloysite was formed by hydrothermal alteration of rhyolitic flow rocks on which residual weathering has been superimposed (Harvey and Murray, 1993). The alteration occurs in a complex system of fractured coalescing rhyolite domes. The altered halloysitic rocks contain about 50% quartz and opal C-T and 50% halloysite with minor amounts of kaolinite. The $<2\,\mu m$ fraction contains a very high content of halloysite, which is the commercial product. The hydrothermal alteration that produced the halloysite took place in Pliocene or Pleistocene time. The processed halloysite is exported worldwide because of the whiteness and translucency it provides to fine china ceramic products. Currently, the Dragon Halloysite Mine in Utah is being reopened.

2. BALL CLAYS

The term ball clay has no mineralogical significance and is a term used to describe fine-grained, highly plastic, sedimentary kaolinitic clay. Burst and Hughes (1994) defined ball clay as a plastic, kaolinitic clay with minor to abundant organic matter producing high green strength and fired strength in

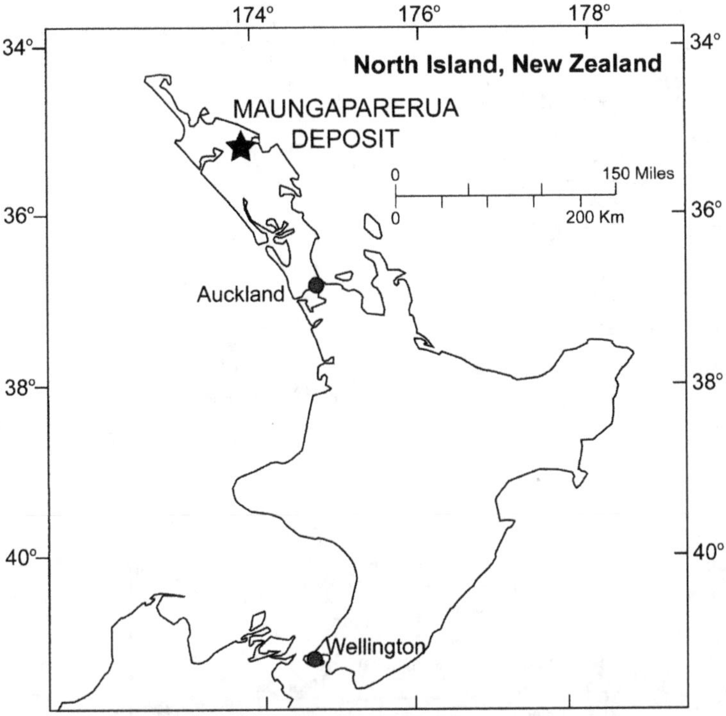

Fig. 34. Location of New Zealand halloysite deposit at Maungaparerua.

ceramic bodies and usually fires white or off-white. The term ball clay originated in England when miners would roll the highly plastic clays into balls that weighed approximately 30–50 lb (Hosterman, 1984). These balls would then be loaded on wagons and transported to the processing plant.

The color of ball clays ranges from light gray to nearly black. Kaolinite is the principal mineral in ball clays, but other minerals are present including illite, smectite, quartz, plagioclase and potash feldspar, and, occasionally, calcite. Organic matter in the form of leaf imprints and disseminated lignitic and platy materials are common in most ball clays. The most important ball clay deposits are located in the United States, England, Germany, and Ukraine. Deposits of ball clay are lenticular bodies thought to have been deposited in swamps, lakes, and oxbows on upper deltaic plains and on river flood plains.

2.1. United States

The largest ball clay producing area is located in western Kentucky and Tennessee (Fig. 35). The age of most of the deposits is Middle Eocene although a few are Lower Eocene (Olive and Finch, 1969). Most deposits

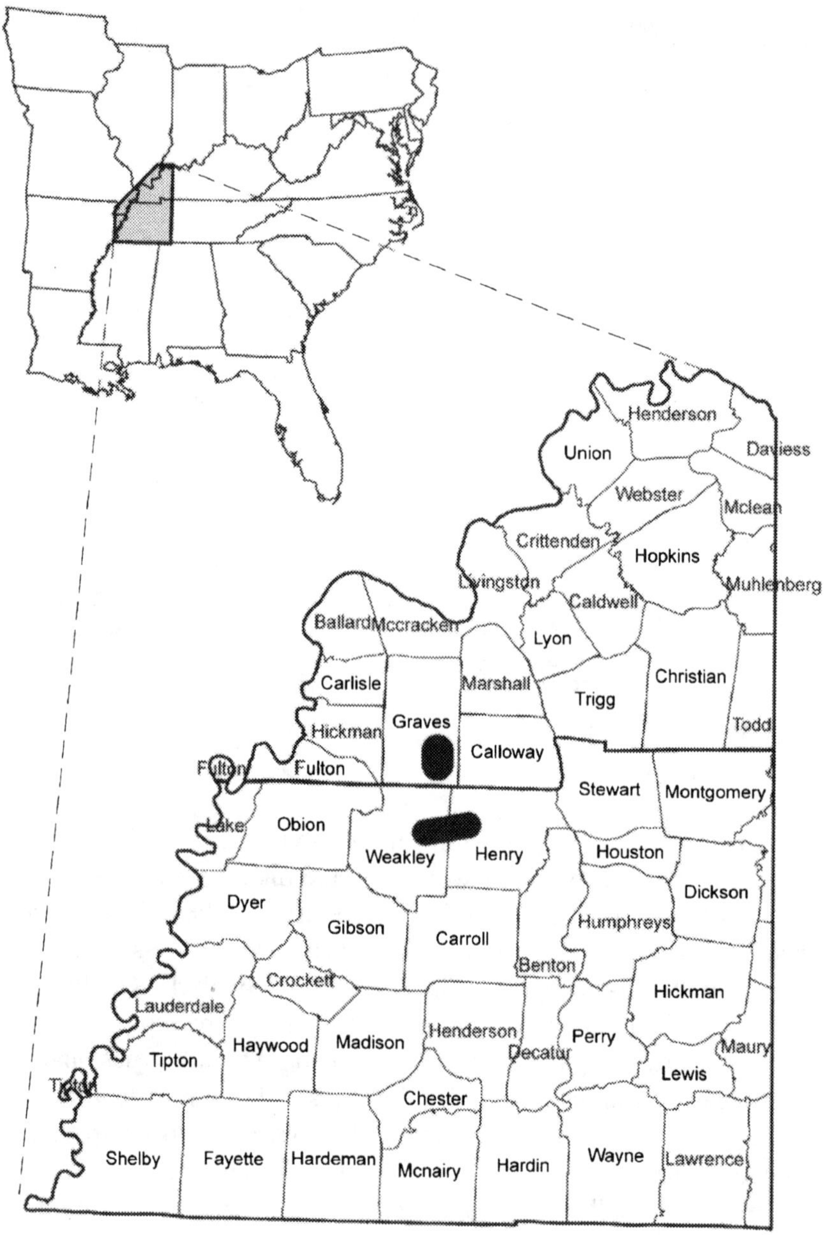

Fig. 35. Location of the ball clay deposits in western Kentucky and Tennessee.

are sedimentary lenticular bodies that vary considerably in size and shape. The thickness of the deposits ranges from 2 to 10 m (McCuiston, 1995). Fig. 36 is a scanning electron micrograph of a Tennessee ball clay, which shows a typical swirl texture characteristic of scanning electron micrographs of ball clay.

Fig. 36. Scanning electron micrograph of a Tennessee ball clay.

Other ball clay areas in the United States from which some ball clay is produced are in Panola County, Magnolia State (Bicker, 1970) and in Texas in Cherokee County (Fisher et al., 1965).

2.2. England

The principal ball clay deposits in England occur in the Bovey Basin southeast of Dartmoor in Devonshire; the vicinity of Petrockstow in North Devonshire; and in the Wareham and Poole regions of Dorsetshire. These ball clays occur in sedimentary rocks of Tertiary age consisting of lenticular units of sand, clay, and lignite of probable lacustrine origin (Scott, 1929). The ball clay is very fine grained and plastic and ranges in color from cream to dark gray or brown. The darker color is caused by carbonaceous matter. The deposits in England and in the Kentucky–Tennessee area are very similar in mineralogy and physical properties.

2.3. Germany

The largest ball clay producing district is the Westerwald area located north and east of Koblenz on the Rhine River. These fine-grained plastic clays are Tertiary in age. They were deposited in subsidence depressions in Devonian age rocks. Most of the ball clay mined is supplied to the ceramic industry (Willis, 2002). The largest quantity is exported to Italy for use in ceramic tile. These Westerwald clays are also used in the manufacture of heavy clay products including brick and tile. The reserves of this ball clay are estimated to last 80–100 years. Currently, around 4.5 million tons are produced annually (Saller, 1999).

2.4. Ukraine

A relatively new source of ball clays is located in eastern Ukraine about 70 km north of Donetsk. Over 1 million tons of ball clay are currently being produced (O'Driscoll, 1998). The age of the ball clays is Tertiary and the thickness is up to 4 m. The mineral content ranges from 55% to 70% kaolinite, 25% to 30% illite, and 5% to 10% quartz. These ball clays burn white or near-white because the iron content is low, ranging from 0.6% to 0.8%. They are very plastic and have a high modulus of rupture.

3. REFRACTORY CLAYS

Refractory clays are comprised largely of kaolinite, which are used to make heat-resistant firebricks, insulating bricks, saggers, refractory mortars and mixes, monolithic and castable materials, ramming and airgun mixes, and other heat-resistant products. As resistance to heat is the most essential property, many specifications include references to pyrometric cones which will be discussed in Chapter 5. Refractory clays consist of flint clays, fireclays, and bauxitic clays. The physical characteristics of fireclay vary considerably ranging from those which are soft and plastic to dense flint-like clays. Flint clay has unique properties for a clay in that it lacks plasticity, breaks with a conchoidal fracture, and is very brittle.

3.1. United States

Most fireclays are of Pennsylvanian age and are widely scattered from Virginia to Missouri (Fig. 37). The fireclays occur below coal beds and are referred to as underclays or seat earths. Opinions differ considerably on the origin of these extensive fireclays occurring below coal seams. Most geologists believe that these underclays formed by the alteration of aluminous sediments in a swamp environment (Patterson and Hosterman, 1963; Keller, 1970). Others believe that these clays were transported and winnowing processes formed rather pure accumulations of kaolinite. High

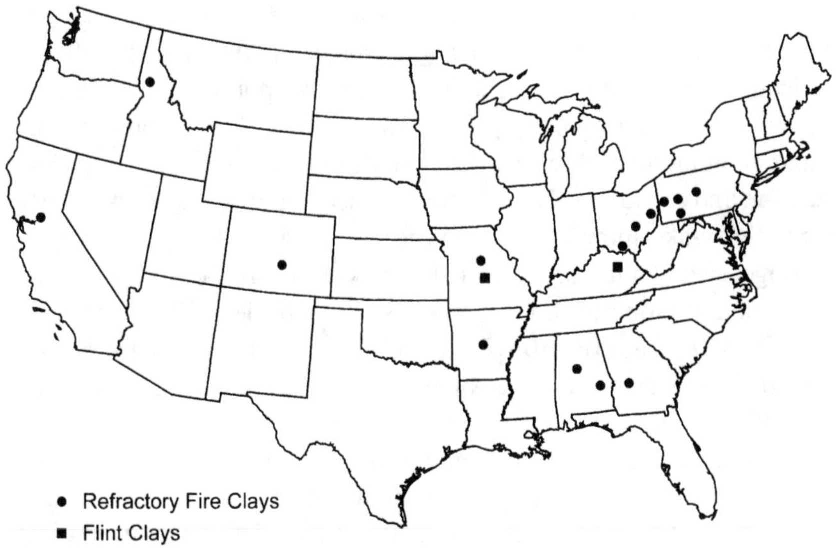

Fig. 37. Location of fireclay deposits including underclays and flint clays.

grade flint clays occur in the Olive Hill District in Kentucky (Patterson and Hosterman, 1963); in the Oak Hill District in Ohio (Stout et al., 1923); and in the Somerset District in Pennsylvania (Hosterman et al., 1968). High grade flint clays are found in Missouri in the Ozark region (Keller et al., 1954). The locations of these flint clay deposits in Kentucky and Missouri are shown in Fig. 37.

Some fireclays in the western United States are Cretaceous in age and are produced in Colorado. Refractory kaolin is produced in Georgia, South Carolina, Alabama, and Idaho. Bauxitic kaolins are located near Andersonville, Georgia and Eufaula, Alabama. Burley clays and diaspore clays are located in Missouri and western Pennsylvania.

3.2. Other Countries

Refractory clays are produced in many countries, including England, Germany, France, Japan, Argentina, Australia, India, Italy, Mexico, Czech Republic, China, Russia, Poland, and Hungary.

4. SMECTITES

Smectite minerals are the dominant component in bentonite, which is a rock term. Many geologists early in the 20th century recognized that bentonite originated from transported volcanic materials. Ross and Shannon (1926) defined bentonite as follows: "Bentonite is a rock composed essentially of a crystalline clay-like mineral formed by devitrification and the accompanying chemical alteration of a glassy igneous material, usually a tuff or volcanic ash; and it often contains variable proportions of accessory crystal grains that were originally phenocrysts in the volcanic glass. These are feldspar (commonly orthoclase and oligoclase), biotite, quartz, pyroxenes, zircon, and various other minerals typical of volcanic rocks. The characteristic clay-like mineral has a micaceous habit and facile cleavage, high birefringence and a texture inherited from volcanic tuff or ash, and it is usually the mineral montmorillonite, but less often, beidellite."

The difficulty in using the above definition for bentonite as an industrial mineral commodity is that it is based on origin and is restricted to volcanic ash, tuff, or glass parent material. Deposits consisting of smectite having other origins or parent materials cannot properly be called bentonite. Perhaps the best definition of bentonite as an industrial mineral is one proposed by Grim and Guven (1978) who defined bentonite as any smectite clay that does not include the mode of origin. This definition solves the problem of the difference between the geologic and industrial usages of the term and overcomes the difficulty in assigning a name to smectite clays that formed from igneous rock other than volcanic or sedimentary or other uncertain origin.

4.1. Western United States Bentonites (South Dakota, Wyoming, and Montana)

The principal producing area of sodium bentonite in the world is from the states of Wyoming, Montana, and South Dakota (Fig. 38). The bentonite beds occur in the New Castle, Mowry, and Belle Fourche Formations of Upper Cretaceous age (Knechtel and Patterson, 1962). There are several bentonite beds (Fig. 39) which are mined but the most important is the clay spur bed which ranges in thickness from about 50 cm to 2 m (Grim and Guven, 1978).

The clay spur bentonite was altered from rhyolitic ash which came from volcanoes located west of the Mowry Seaway into which the ash fell (Slaughter and Early, 1965). The trace element chemistry of the sodium montmorillonite separated from the clay spur bentonite provides strong evidence that the ash altered while in contact with the Mowry seawater (Elzea and Murray, 1990). The clay spur bentonite is composed of 90% or more sodium montmorillonite along with minor quantities of quartz, feldspar, biotite, opal C-T, and apatite.

The color of the clay spur bentonite is olive green or yellow when oxidized by weathering and is tan to blue gray when unweathered. In

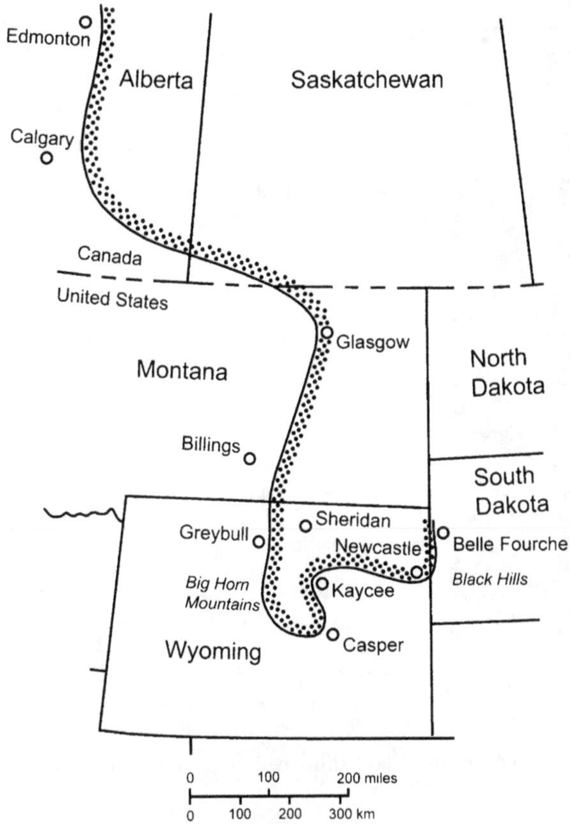

Fig. 38. Location of outcrop area of the Mowry sodium bentonites.

general, the weathered and oxidized bentonite has higher viscosity and gel strength and a lower filtrate loss than the unweathered bentonite (Elzea and Murray, 1990). Because the weathered and oxidized bentonite has superior colloidal properties, most of the bentonite is stockpiled for several months before it is processed in order to let it weather and oxidize. The western sodium bentonites are high swelling, viscous, and thixotropic clays. These Cretaceous deposits are high quality and are utilized around the world in drilling fluids and foundry mixes. Fig. 11 is an electron micrograph of a sodium bentonite from Wyoming showing the typical cornflake appearance.

4.2. Southern United States Bentonites (Texas, Mississippi, and Alabama)

The bentonites which occur in the Gulf Coast area are generally referred to as southern or non-swelling bentonites. These bentonites are composed

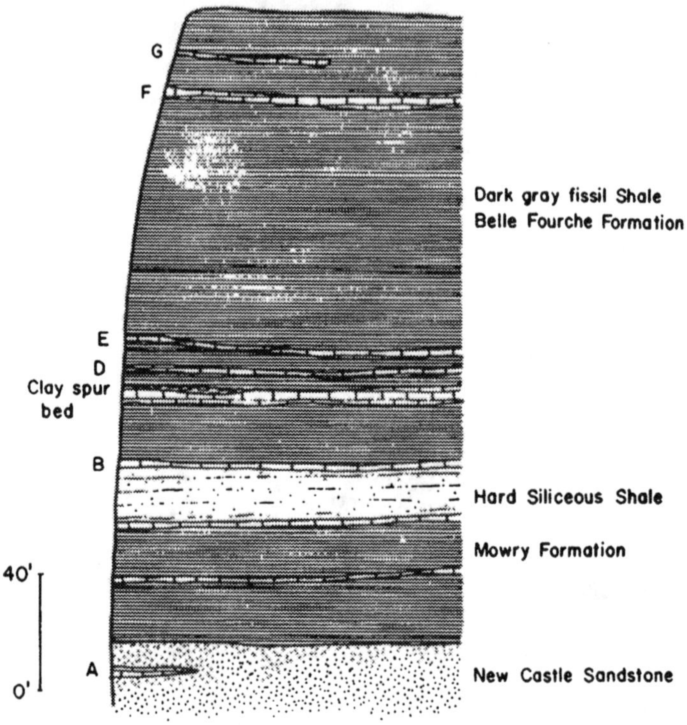

Fig. 39. Mowry bentonite layers including the clay spur bed and other bentonite layers A–G.

primarily of calcium montmorillonite. The major deposits are located (Fig. 40) in Texas, Mississippi, and Alabama in formations ranging in age from Upper Cretaceous to Middle Tertiary (Grim and Guven, 1978).

In Texas, the bentonites are best developed in the Tertiary Jackson and Gueydan Formations in Gonzalez and Lafayette counties. Hagner (1939) pointed out that the source of the volcanic ash was southwest of the bentonite deposits. The volcanic ash was deposited over a considerable area and the devitrification to form the bentonites took place in both marine and lacustrine environments (Roberson, 1964). Roberson classified the crystallinity of the montmorillonites in the Jackson and Gueydan bentonite as either well crystallized or poorly crystallized and suggested that the well crystallized montmorillonites formed in place and the poorly crystallized montmorillonites were redeposited. The color of these Texas bentonites varies from chocolate brown to olive to white. The thickness varies from about 1 to 3 m. The white bentonite in the Jackson Formation is an altered airborne vitric dust or a bed of volcanic vitric claystone (Chen, 1970).

Bentonite is produced in Itawanba and Monroe counties in Mississippi from the Upper Cretaceous Eutaw Formation (Grim and Guven, 1978).

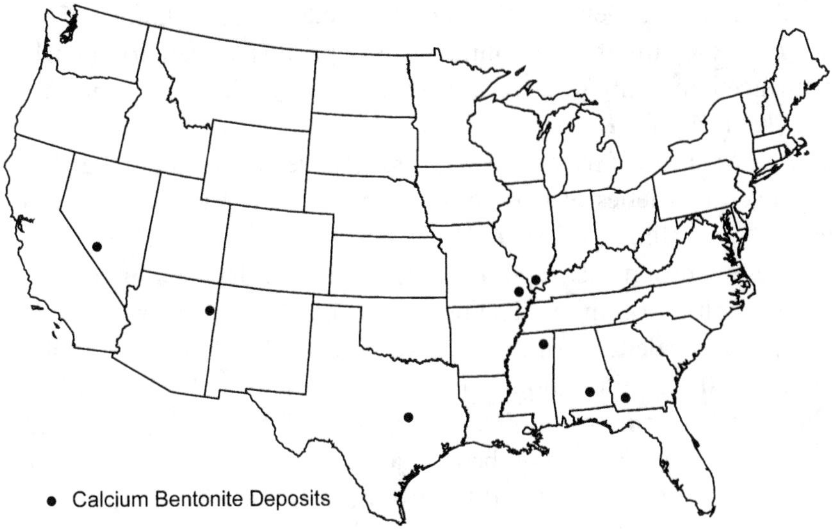

Fig. 40. Location of calcium bentonite deposits in Alabama, Arizona, Georgia, Illinois, Mississippi, Missouri, Nevada, and Texas.

Some of these bentonite beds reach a thickness of 14 ft. The material is waxy blue in color when fresh and is yellow when weathered. The Mississippi bentonites are calcium, magnesium montmorillonites and contain quartz, feldspar, and mica impurities. The source of the volcanic ash from which these Upper Cretaceous bentonites formed has not been determined although a probable source was the Midnight volcanic system near Jackson, Mississippi, which erupted in Cretaceous time (Merrill, 1983).

In Alabama, bentonite is mined and produced from the Ripley Formation of Paleocene age at Sandy Ridge.

4.3. Southwestern Bentonites (Arizona, California, and Nevada)

A white bentonite at Cheto, Arizona (Fig. 40) is comprised mainly of calcium montmorillonite. At this location, a 20 ft bed of ash of latitic composition in the Pliocene Bidahochi Formation has altered to calcium montmorillonite (Grim and Guven, 1978). This Arizona bentonite has a high magnesium content in the range of 5–6%. Impurities present in this bentonite are quartz, mica, feldspar, and some kaolinite. The source of the vitric ash which was the precursor of the Cheto bentonite is a matter of speculation because of the extensive volcanic activity during Pliocene time (Kiersch and Keller, 1955). Volcanic activity took place to the west, northeast and east, and to the south, so that the ash could have come from any one of these sources and been deposited in depositional traps in close proximity to the Defiance Plateau. The beds of the Bidahochi Formation which contain the bentonite were deposited on an irregularly eroded surface consisting of stream channels, lake basins, and other depressions (Kiersch and Keller, 1955).

In the Ash Meadows area of the Amargosa Valley of Nevada (Fig. 40), there are several varieties of bentonites ranging in age from Miocene to Pleistocene. According to Papke (1969) the dominant component of many of these bentonites is saponite. There are also deposits of sodium and calcium montmorillonites. The bentonites contain considerable mica and dolomite and calcite (Grim and Guven, 1978). The Ewing bentonite consists of 80–90% sodium montmorillonite. The Kinney bentonite contains about 90% montmorillonite, which is a mixture of sodium and calcium types. The playa lake beds that contain the bentonites are Pleistocene in age and are believed to have been formed by alteration of tuffaceous sediments (Papke, 1969).

Near Hector, California in Central San Bernadino County, is an occurrence of a lithium montmorillonite called hectorite. The hectorite deposit formed by the reaction of lithium- and fluorine-bearing hot water which rose along a fault zone which reacted with travertine to produce a high magnesia, low silica, swelling bentonite containing 1% lithium (Hosterman and Patterson, 1992). Another hectorite occurrence is in northern Nevada.

4.4. *Fuller's Earth Deposits (Georgia, Illinois, Mississippi, Missouri, and Tennessee)*

Fuller's earth is defined as any clay which has a high absorbent capacity and/or are natural bleaching earths. Therefore, the term has no genetic significance. These clays have a high content of smectite and many are classed as bentonites according to the definition prescribed by Grim and Guven (1978).

The absorptive clays that are mined in Missouri, Illinois, Tennessee, and Mississippi are in the Paleocene Porters Creek Formation (Fig. 41). These deposits are located along the margin of the Mississippi Embayment near Bloomfield, Missouri; Olmstead, Illinois; Paris and Middleton, Tennessee; and Ripley, Mississippi (Fig. 42). The Porters Creek ranges in thickness from 15 to 25 m and is dark gray in color, which with prolonged exposure to weathering turns beige. The major mineral present is a calcium montmorillonite along with some illite and kaolinite and mixed layer illite–smectite. Non-clay minerals present in minor quantities are quartz, feldspar, opal C-T, clinoptilolite, and siderite. The sources of the clays that make up the Porters Creek were the smectite Cretaceous clays

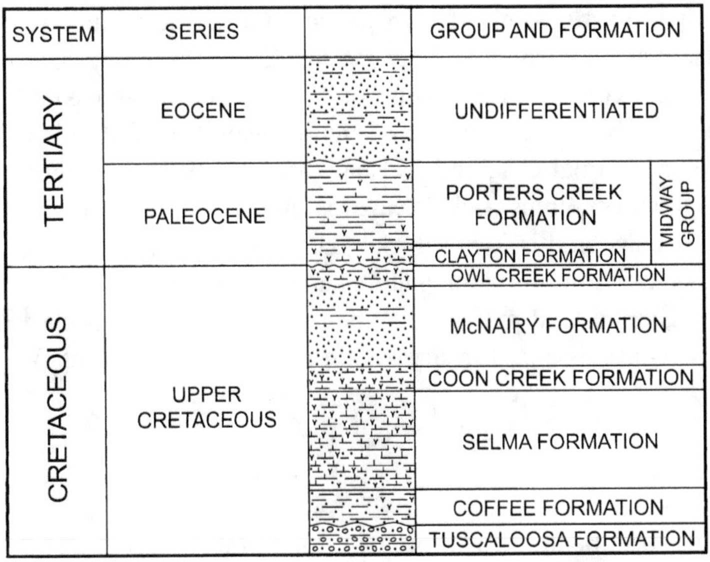

Fig. 41. Sorptive clays occur in the Paleocene Porters Creek Formation.

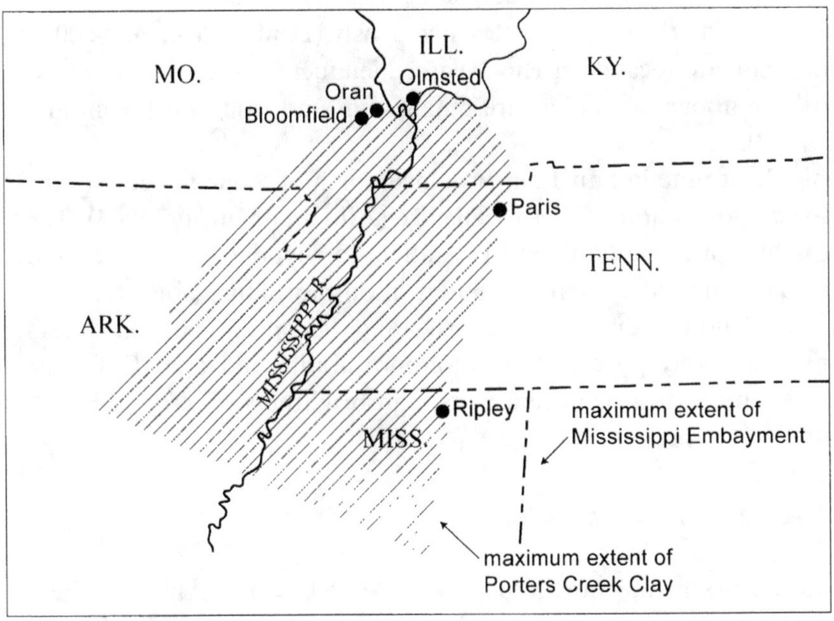

Fig. 42. Location of plants that produce Porters Creek sorptive clays.

from the Western Interior and the illitic and kaolinitic clays from the Piedmont (Thomas and Murray, 1989).

Two types of fuller's earth deposits are found in Georgia—the mixed smectite—palygorskite clay in the Ochlocknee-Meigs area of South Georgia and the Twiggs clay in East Georgia. The mixed smectite–palygorskite clay is found in the Hawthorne Group of Miocene age. This absorbent clay, which is 10–15 m thick, consists of a mixture of smectite and palygorskite, along with quartz and abundant diatoms and minor quantities of 10–15 m thick kaolinite, dolomite, opal C-T, and feldspar (Merkl, 1989). In eastern Georgia, near the town of Wrens, fuller's earth is mined from the Twiggs clay of Upper Eocene age and is 3–5 m thick. The minerals present in this Twiggs clay are calcium, montmorillonite, illite, kaolinite, quartz, opal C-T, calcite, hematite, aragonite, and marcasite (Weng, 1995). The Twiggs clay occurs above the Cretaceous and Tertiary kaolin and is exposed in many kaolin mines.

4.5. Argentina Bentonites

Bordas (1943) described bentonites located in Mendoza and San Juan Provinces of Triassic age and in Paleocene and Eocene Formations in Patagonia. The bentonites in Mendoza and San Juan Province are altered tuffs interbedded with sand and conglomerates. The Patagonia bentonites are frequently interbedded with unaltered ash (Lombardi et al., 2003). Other deposits are located in Rio Negro, Neuquen, and San Juan Provinces. All the above bentonites are magnesium and iron rich (Grim and Guven, 1978).

A white bentonite in San Juan Province north of Rodeo occurs in the Las Trancas Formation of Pleistocene age (Allo and Murray, 2004). The deposit is hydrothermally altered rhyolitic and rhyodacitic pumices and breccias. The mineral content is sodium montmorillonite, biotite, muscovite, clinoptilolite, opal C-T, quartz, and feldspar. The –325 mesh fraction is relatively pure sodium montmorillonite and opal C-T. The unusual feature of this bentonite is its high brightness of about 85% or more after wet processing.

4.6. Africa (Morocco and Algeria)

Bentonite occurs in the northern part of Morocco and Algeria. These bentonites are Cretaceous and Tertiary in age and are the product of altered ash beds and altered rhyolite (Grim and Guven, 1978). The bentonite near Camp Berteaux of Miocene age is used for decolorizing oils. These bentonites are magnesium rich and the extractable interlayer cations are magnesium and sodium, except for the Camp Berteaux bentonite in which the extractable cations are magnesium, calcium, and sodium (Grim and Guven, 1978).

4.7. Union of South Africa and Mozambique

The principal bentonites mined in South Africa are in the vicinity of Plettensberg Bay and Albertinia. These bentonites are Tertiary in age and are relatively pure montmorillonite about 2 m thick.

In Mozambique, about 40 km east of Lorenco Marques, a perlitic lava of the Stormberg Series of the Karoo System of Liassic age has been altered by deuteric action to bentonite. Some of the bentonite are tens of meters thick. It is a calcium montmorillonite (Grim and Guven, 1978), which is used primarily by the foundry industry.

5. EUROPEAN BENTONITES

5.1. England

The most important bentonite in England is located near Redhill in Surrey. This bentonite (called fuller's earth in England, Robertson, 1986) is Cretaceous in age and is a calcium montmorillonite. This bentonite is used in many industrial applications including bonding foundry molding sands, drilling muds, acid-activated decolorizing clays, cat litter, and many other miscellaneous uses. A stratigraphic section of this bentonite is shown in Grim and Guven (1978). Another calcium bentonite of Jurassic age is located in Somerset (Hallam and Selwood, 1968).

5.2. Germany

Bentonite deposits near Moosburg and Lanshut in Bavaria are and have been important commercial bentonites for European and world markets. The bentonites are Upper Miocene in age and are relatively pure calcium montmorillonites. These bentonites are altered acid vitreous tuffs (Grim and Guven, 1978). The deposits are rather small areally and some are as much as 3 m thick. These bentonites are used as acid-activated decolorizing clays, and after treatment with soda ash or sodium polyacrylate, are used for drilling muds, bonding molding sands in foundries, and barrier clays in environmental applications.

5.3. Greece

The best-known and most widely used bentonites in Greece are located on the Island of Milos. These bentonites of Pleistocene age were formed from pyroclastic flows (Christidis, 2001). Both deuteric and hydrothermal activities have been important in the formation of the bentonite. The Milos bentonites are primarily calcium montmorillonites and have a relatively high magnesium content (Grim and Guven, 1978). The Milos bentonite is exported for use in iron ore pelletizing, cat litter, and for bonding foundry molding sands. The deposits are quite thick, exceeding 50 m in places, but are not laterally extensive. The commercial deposits are located in eastern Milos.

5.4. *Italy*

Important calcium bentonite deposits are located on the Island of Sardinia (Carta et al. 1977). There are two types of deposits on Sardinia, a sedimentary bentonite of Miocene age and hydrothermally altered trachytic tuffs (Annedda, 1956). The sedimentary bentonite is a marine altered volcanic ash. Another bentonite deposit in Sardinia was described by Pietracaprina et al. (1987).

6. ASIA

6.1. *China*

Bentonite production in China in 2001 was about 2 million tons (Crossley, 2003). The largest deposits are in Zhejiang, Hubei, Anhui, Shandsing, Sichuan, and Liuoning Provinces. These bentonites are mainly calcium montmorillonites which are treated with soda ash to convert some to a sodium montmorillonite. The reserves of bentonite in China are estimated to be 2,500,000,000 tons. The growth of bentonite production in China is over 7% annually.

6.2. *India*

The major bentonite mines in India are located in Kutch with processing plants in Bhuj and Chennar (Crossley, 2003). Bentonite is also produced in the Barmer District of Rajasthan near Akliand Hathi-ki-Dhani (Siddiquie and Bahl, 1965). These deposits are Lower Tertiary in age and are relatively pure calcium montmorillonite up to 3 m thick.

6.3. *Japan*

Bentonites of Miocene and Pliocene age formed by alteration of volcanic ash, pumice, and tuff are located in Yamagata, Gumma, and Nagano (Sudo and Shimoda, 1978) Prefectures on Honshu and Hokkaido Islands.

These bentonites are sodium bentonites with cation exchange capacity values in the mid-1980s. The major impurities are quartz, opal C-T, iron oxides, and in Yamagata sericite. Some of the bentonites in the Gumma Prefecture are up to 5 m thick (Iwao, 1969).

6.4. Georgia Republic

The most well-known bentonite deposit in the Georgia Republic is in the Askana area. These bentonites were formed by the alteration of andesite–trachyte tuffs of Eocene age (Grim and Guven, 1978). Both sodium and calcium montmorillonites are mined in the Askana deposits. The major impurities are opal C-T and quartz. Rateev (1967) believes the Askana bentonites are hydrothermal. Another important bentonite is the Gumbri deposit which is Cretaceous in age (Dsotsenidze and Matchabely, 1963). Some of the Gumbri bentonite are 8 m thick. This bentonite was formed by submarine alteration of volcanic glass.

7. PALYGORSKITE (ATTAPULGITE) AND SEPIOLITE

7.1. United States

The palygorskite sedimentary deposits in North Florida and South Georgia (Fig. 43) are Middle Miocene in age and are in the Hawthorne Formation (Merkl, 1989; Krekeler, 2004). The depositional environment was estuaries and marine lagoons which varied from saline to nearly fresh water in a shallow low energy environment (Merkl, 1989; Krekeler et al., 2004). The major mineral present is palygorskite along with smectite, illite, quartz, dolomite, apatite, iron oxides, anatase, and cassiterite (Krekeler et al., 2004). The thickness of these palygorskite clays is about 2–3 m. A small sepiolite deposit is located in the Ash Meadows area in Nevada about 100 miles north of Las Vegas. This deposit is about 1 m in thickness and occurs as stringers in saponite which is Pliocene or Plastocene in age (Khoury et al., 1982).

7.2. China

Palygorskite is mined near the provincial boundary between Anhui and Jiangsu Provinces (Fig. 44). The deposits are Middle Miocene in age and occur in the Huaguoshan Formation. The deposits are the alteration product of a basaltic ash (Zhou and Murray, 2003). The Geological Survey of China estimated that there are over 200 million tons of reserves. The

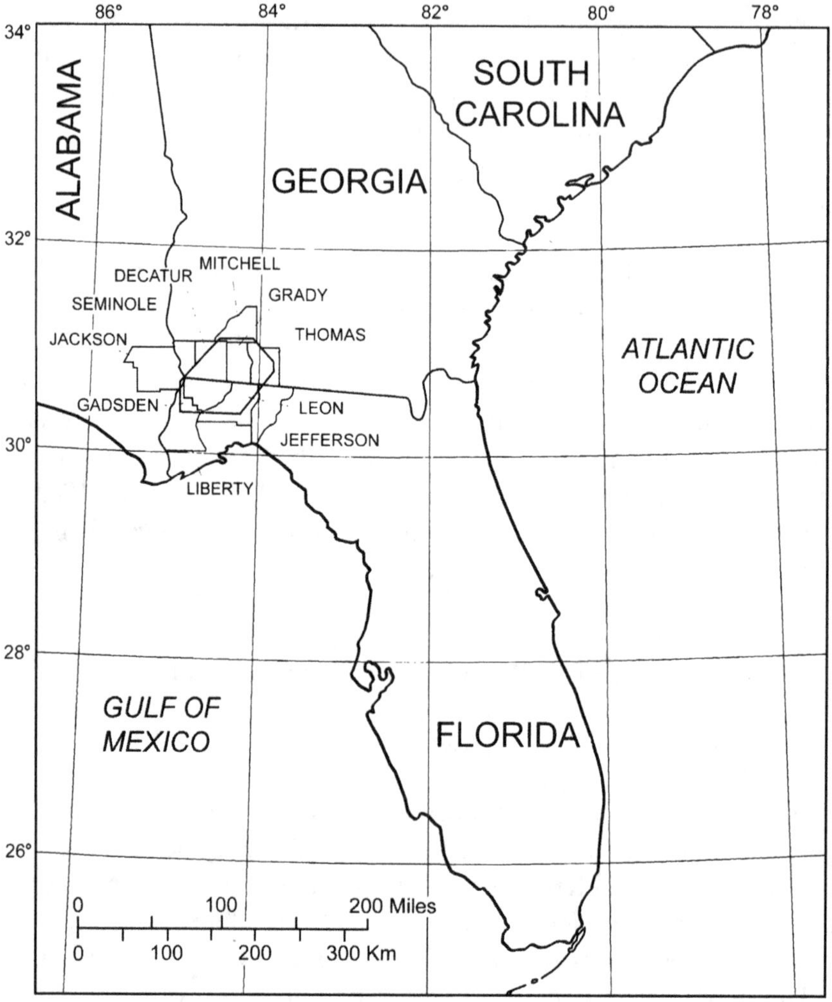

Fig. 43. Location of palygorskite deposits in southern Georgia and northern Florida.

deposits range in thickness between 3 and 6 m. The dominant mineral is palygorskite with minor amounts of smectite, quartz, and dolomite. At Mingguang in Anhui Province, the palygorskite occurs immediately under a basalt. In the vicinity of Xuyi in Jiangsu Province, the palygorskite is very near the surface and is overlain by a thin smectite layer.

7.3. Senegal

Palygorskite is mined near the town of Theis, which is about 100 km east of Dakar. The palygorskite overlies an aluminum phosphate deposit

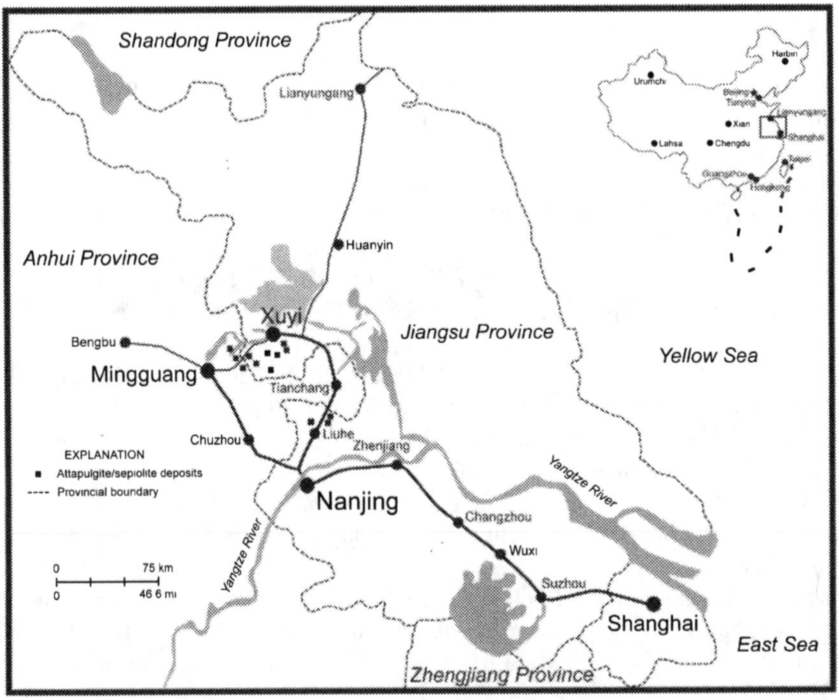

Fig. 44. Location of palygorskite deposits in central China.

which is also mined. The palygorskite is Early Eocene in age and ranges from 2 to 6 m in thickness (Wirth, 1968). The major mineral present is palygorskite, along with minor amounts of quartz, dolomite, chert, and sepiolite. The palygorskite beds extend south–southwest from Theis to the southern border of Senegal, a distance of about 100 km.The palygorskite is mined, processed, and hauled to Dakar, where it is loaded onto ocean going vessels and shipped mainly to Europe.

7.4. Spain

The largest commercial sepiolite deposit in the world is at Vicalvaro near Madrid. The sepiolite was precipitated in an evaporate sequence in Tertiary lakes (Huertas et al., 1971). The thickness ranges from 1 to 5 m. Sepiolite and palygorskite are produced at Toledo, Torrejon el Ruhe, Lebija, Almeria, and Maderuelo (Griffiths, 1991). Galan et al. (1975) described a palygorskite deposit in the province of Caceres in west central Spain. Spain is the largest producer of sepiolite clays in the world.

REFERENCES

Allo, W. and Murray, H.H. (2004) Mineralogy, chemistry and potential applications of a white bentonite in San Juan Province, Argentina. *Appl. Clay Sci.*, 25, 237–243.

Alves, Carlos Alberto, Personal communication.

Annedda, V. (1956) Deposits of bentonite on Sadali and Vallanova Tulo Territories, Sardinia. *Resoconti Assoc. Min. Sarda*, 60, 5–9.

Aron, L. (2000) Excursion guide. *Proceedings of the XVIth Conference on Clay Mineralogy and Petrology*. Czech National Clay Group, Karlovy Vary, Czech Republic.

Bicker, A.R. (1970) *Economic Minerals of Mississippi*. Bulletin 112, Mississippi Geological, Economical and Topographical Survey, 80pp.

Bordas, A.F. (1943) Contribucion al conocimientas de las bentonites Argentinas. *Rev. Mineral. Geol. Mineral.*, 14, 1–60.

Bristow, C.M. (1993) *The Genesis of the China Clays of South-West England—A Multistage Story*. Special Publication No. 1. Clay Minerals Society, Boulder, CO, pp. 171–203.

Burst, J.F. and Hughes, R.E. (1994) Construction uses: clay based ceramic raw materials. Chapter in *Industrial Minerals and Rocks*, 6th Edition. Carr, D.D., ed. Society for Mining, Metallurgy and Exploration, Littleton, CO, pp. 317–324.

Carta, M.C., Ciccu, R., and Pretti, S. (1977) The industrial minerals of Sardinia: present situation and future perspectives. *Proceedings of the 2nd Industrial Minerals International Congress*, Munich, pp. 41–55.

Chen, P.Y. (1970) Petrography of the Upper Eocene ash beds from Gonzales and Fuyette Counties, Texas. *Proc. Geol. Soc. China*, 13, 23–33.

Christidis, G.E. (2001) Geochemical correlation of bentonites from Milos Island, Aegean, Greece. *Clay Miner.*, 36, 295–306.

Crossley, P. (2003) Bent on success. *Ind. Miner.*, June, 32–37.

Dewu, B.B.M. and Durrance, E.M. (1993) *Mobility of U and Granite Kaolinization in Southwest England*. Special Publication No. 1. Clay Minerals Society, Boulder, CO, pp. 205–219.

Dombrowski, T. (1993) *Theories of Origin for the Georgia Kaolins*. Special Publication No. 1. Clay Minerals Society, Boulder, CO, pp. 75–97.

Dsotsenidze, G.S. and Matchabely, G.A. (1963) The genesis of bentonites of Georgian SSR. *Proceedings of the International Clay Conference*, Stockholm 2, pp. 203–210.

Elzea, J. and Murray, H.H. (1990) Variation in the mineralogical, chemical, and physical properties of the Cretaceous clay spur bentonite in Wyoming and Montana. *Appl. Clay Sci.*, 5, 229–248.

Fisher, W.L., et al. (1965) *Rock and Mineral Resources of East Texas*. Bureau of Economic Geology, University of Texas, Report of Investigations 54, 439pp.

Franca, M. (2002) Brazilian coating kaolins. *Society for Mining, Metallurgy and Exploration—Dreyer Conference*, September 2002, Savannah, Georgia.

Galan, E., et al. (1975) The Caceres palygorskite deposit. *Proceedings of the International Clay Conference*, Mexico City, Mexico, pp. 81–94.

Griffiths, J. (1991) *Spain's Minerals: Mixed Fortunes*. Industrial Minerals, No. 285, London, pp. 23–47.

Grim, R.E. and Guven, N. (1978) *Bentonites—Geology, Mineralogy, Properties and Uses*. Elsevier, Amsterdam, 256pp.

Hagner, A.F. (1939) Adsorptive clays of the Texas Gulf Coast. *Am. Mineral.*, 24, 67–108.

Hallam, A. and Selwood, B.W. (1968) Montmorillonite and zeolite in Mesozoic and Tertiary beds of Southern England. *Mineral. Mag.*, 37, 945–952.

Harvey, C.C. and Murray, H.H. (1993) *The Geology, Mineralogy and Exploitation of Halloysite Clays of Northland, New Zealand*. Special Publication No. 1. Clay Minerals Society, Boulder, CO, pp. 233–248.

Hinckley, D.N. (1963) Variability in crystallinity values among the kaolin deposits of the Coastal Plain of Georgia and South Carolina. *Clay. Clay Miner.*, 11, 229–235.

Hosterman, J.W. (1984) *Ball Clay and Bentonite Deposits of the Central and Western Gulf of Mexico Coastal Plains, United States*. US Geological Survey Bulletin 1558-C, p. 22.

Hosterman, J.W. and Patterson, S.H. (1992) *Bentonite and Fuller's Earth Resources of the United States*. US Geological Survey, Professional Paper 1522, 45pp.

Hosterman, J.W., et al. (1968) *Clays*. US Geological Survey, Professional Paper 580, pp. 167–188.

Huertas, F.J., et al. (1971) Minerales Fibrosos de la Arcilla en Cuencas Sedimentorias Espanolas 1 Cuenen del Tajo. *Bul. Geol. Y. Min.*, 82, 28–36.

Iwao, S. (1969) The Clays of Japan. Geological Survey of Japan. *Proceedings of the International Clay Conference*, University of Tokyo, Tokyo, Japan, 209pp.

Keller, W.D. (1970) Environmental aspects of clay minerals. *J. Sediment. Petrol.*, 40, 788–813.

Keller, W.D., et al. (1954) The origin of Missouri fire clays. Clays and clay minerals. *Proceedings of the 2nd National Conference on Clay Minerals*, 1953, Swineford, A., ed. US National Research Council Publication 327, pp. 7–46.

Khoury, H.N., et al. (1982) Origin of magnesium clay from the Amargosa Desert, Nevada. *Clay. Clay Miner.*, 30, 327–336.

Kiersch, G.A. and Keller, W.D. (1955) Bleaching clay deposits. Chapter in *Mineral Resources, Navajo-Hopi Reservation*, Vol. 2. Kiersch, G.A., ed. Nonmetallic Minerals, Geology, Evaluation and Uses. University of Arizona Press, Tucson, AZ, pp. 41–54.

Knechtel, M.M. and Patterson, S.H. (1962) *Bentonite Deposits of the Northern Black Hills District, Wyoming, Montana and South Dakota*. US Geological Survey Bulletin 1082, 893–1029.

Kogel, J.E., et al. (2002) *The Georgia Kaolins—Geology and Utilization*. Society for Mining, Metallurgy and Exploration, Littleton, CO, 84pp.

Krekeler, M.P.S. (2004) Improved constrains on sedimentary environments of palygorskite deposits of the Hawthorne Formation, Southern Georgia from a detailed study of a core. *Clay. Clay Miner.*, 52, 253–262.

Krekeler, M.P.S., et al. (2004) A microtexture study of palygorskite-rich sediments from the Hawthorne Formation, Southern Georgia, by transmission electron microscopy and atomic force microscopy. *Clay. Clay Miner.*, 52, 263–274.

Krelina, et al. (1982) *Zavercena Zprava Ukolu Karlovarsko Kaolin a Karlovarsko Reinterpretace (studie)*. MS Geofond, Praha.

Kuzvart, M. (1984) *Industrial Minerals and Rocks*. Elsevier Science Publishing Co., New York, NY, 454pp.

Lin, Y.C., et al. (1989) A tentative discussion on the Maoming kaolin deposit, mineral composition, papermaking technological properties and ore-forming geological conditions. *Miner. Deposit.*, 8, 55–64.

Lombardi, B., et al. (2003) Bentonite deposits of Northern Patagonia. *Appl. Clay Sci.*, 22, 309–312.

McCuiston, J.T. (1995) *The Mineralogy and Physical Characteristics of the Ball Clays of Henry and Weakley Counties, Tennessee*. MS Thesis, Indiana University, 142pp.

Merkl, R.S. (1989) *A Sedimentological, Mineralogical and Geochemical Study of the Fuller's Earth Deposits of the Miocene Hawthorne Group of South Georgia—North Florida*. PhD Dissertation, Indiana University, 182pp.

Merrill, R.K. (1983) Source of the volcanic precursor to Upper Cretaceous bentonite in Monroe County, Mississippi. *Mississippi Geol.*, 3, 1–6.

Murray, H.H. (1976) The Georgia sedimentary kaolins. *Proceedings of the International Geological Correlation Program, Project 23, 7th Symposium on Genesis of Kaolin*, Tokyo, Japan, pp. 114–125.

Murray, H.H. (1981) *Kaolin Project in the Amazonia Area, Brazil*. SEM Preprint 81-57, 7pp.

Murray, H.H. and Keller, W.D. (1993) *Kaolins, Kaolins and Kaolins*. Special Publication No. 1. Clay Minerals Society, Boulder, CO, pp. 1–24.

Murray, H.H. and Partridge, P. (1982) Genesis of Rio Jari kaolin. *Proceedings of the International Clay Conference*, 1981, Van Olphen, H. and Veniale, F., eds., Elsevier, Amsterdam, The Netherlands, pp. 279–291.

Murray, H.H., et al. (1978) Alteration of a granite to kaolin—mineralogy and geochemistry. *Schriftenr. Geol. Wiss. (Berlin)*, 11, 197–208.

Neuzil and Konta, J. (1965) Petrology and Geochemistry of the Karlovy Vary Granite, Parent Rock of the Kaolin in Sedlec. *Acta Univ. Carol. Geol. Suppl.*, 2, 51–56.

O'Driscoll, M.J. (1998) Ukraines minerals. *Ind. Miner.*, 373, 21–43.

Olive, W.W. and Finch, W.I. (1969) *Stratigraphic and Mineralogic Relations and Ceramic Properties of Clay Deposits of Eocene Age in the Jackson Purchase Region, Kentucky and Adjacent Parts of Tennessee*. Bulletin 1282, US Geological Survey, 33pp.

Papke, K.G. (1969) Montmorillonite deposits in Nevada. *Clay. Clay Miner.*, 17, 211–222.

Patterson, S.H. and Hosterman, J.W. (1963) *Geology and Refractory Clay Deposits of the Haldeman and Wrigley Quadrangle, Kentucky*. Bulletin 1122-F, US Geological Survey, 113pp.

Pietracaprina, A., et al. (1987) A new bentonite deposit in Sardinia. *Appl. Clay Sci.*, 2, 167–174.

Rateev, M.A. (1967) Sequence of hydrothermal alteration of volcanic rocks into bentonite clays of the Askana deposit in the Georgia SSR. *Dokl. Akad. Nauk. SSSR*, 175(3), 675–678.

Roberson, H.E. (1964) Petrology of Tertiary bentonites of Texas. *J. Sediment. Petrol.*, 43, 401–411.

Robertson, R.H.S. (1986) *Fuller's Earth—A History of Calcium Montmorillonite*. Volturna Press, Hythe, Kent, UK, 421pp.

Ross, C.S. and Shannon, E.V. (1926) Minerals of bentonite and related clays and their physical properties. *J. Am. Ceram. Soc.*, 9, 77–96.

Saller, M. (1999) Germany's minerals reviewed. *Ind. Miner.*, 383, 25–39.

Scott, A. (1929) *Ball Clays*. Special Report, Great Britain Geological Survey, Vol. 31, 73pp.

Siddiquie, N.N. and Bahl, D.P. (1965) *Geology of the Bentonite Deposits of the Barmer District, Rajasthan*. Memoirs of the Geological Survey, India, 36pp.

Slaughter, M. and Early, J.M. (1965). *Mineralogical and Geological Significance of the Mowry Bentonites*. Special Paper, Geological Society of America, Vol. 83, 116pp.

Stout, W.G., et al. (1923) *Coal Formation Clays of Ohio*. Bulletin 26, Ohio Geological Survey, 4th Series, 588pp.

Sudo, T. and Shimoda, S. (1978) *Clays and Clay Minerals of Japan*. Elsevier, Amsterdam, The Netherlands, 326pp.

Thomas, A.R. and Murray, H.H. (1989) Clay mineral segregation by flocculation in the Porters Creek Formation. *Clay. Clay Miner.*, 37, 179–184.

Virta, R.L. (2004) Kaolin. *Ind. Miner. Rev. Min. Eng.*, 56, 28–29.

Weng, X. (1995) *Mineralogy, Morphology, Geochemistry and Selected Physical Properties of the Twiggs Clay in the Georgia Kaolin District*. MS Dissertation, Indiana University, 74pp.

Willis, M. (2002) An eye on the ball. *Ind. Miner.*, October, 26–33.

Wilson, I. (2002) Kaolin. *Mining Jour. Ltd., Annual Miner. Rev.*, p. 6.

Wirth, L. (1968) *Attapulgites du Senegal Occidental*. Rapport No. 26, Laboratoire de Geologie, Faculté des Sciences, Université de Dakar, 55pp.

Yuan, J. (1994) *Clay Mineralogy and Its Influence on Industrial Uses of Some Kaolin Clay Deposits from South China and Eastern Washington—Idaho, USA*. PhD Dissertation, Indiana University, 153pp.

Yuan, J. and Murray, H.H. (1993) *Mineralogical and Physical Properties of the Maoming Kaolin from Guandong Province, South China*. Special Publication No. 1. Clay Minerals Society, Boulder, CO, pp. 249–259.

Zheng Zhi, et al. (1982) Kaolin deposits of China. *Proceedings of the International Clay Conference*, 1981, Van Olphen, H. and Veniale, F., eds., Elsevier, Amsterdam, The Netherlands, pp. 719–731.

Zhou, H. and Murray, H.H. (2003) Geology and physical properties of palygorskite from Central China and Southeastern United States. *Proceedings of the 12th International Clay Conference*, Bahia Blanca, Argentina, Dominguez, E., et al., eds., Elsevier, Amsterdam, The Netherlands, pp. 369–375.

Chapter 4

EXPLORATION, MINING, AND PROCESSING

Once a clay deposit has been discovered, a drilling program to determine the areal extent, thickness and type of overburden, and thickness and quality of the clay is initiated. Both core and auger drills are used in exploration and mine development. Some companies use air drills after the overburden is removed for closely spaced information for mine control. In remote and heavily forested areas such as in Brazil, in the early exploration stages, hand dug pits were used to evaluate the kaolin deposits. These test pits were about 1 m in diameter and up to 30 m deep. The reason that the test pits were used was because of the difficulty in getting truck mounted drills into the area. Once a deposit was discovered, then the area was cleared so that truck mounted drills could be used.

For most clay deposits, a grid pattern of drilling is used to determine the size and quality of the deposit. The distance between the drill holes in the grid pattern is based on several factors including topography, geology, lateral extent and uniformity, and water table considerations. For example, the drilling and testing program is very different if the clay deposit is sedimentary or residual from a deposit that is hydrothermal or is structurally deformed by faults, folds, and inclined beds. Professional advice from a geologist or mining engineer is very helpful in designing a drilling and testing program.

Drilling programs used for the relatively flat lying kaolin deposits in Georgia and Brazil are to initially drill at least 25 holes per 100 acres (40 ha). The purpose of a kaolin drilling program is shown in Table 10.

The overburden is drilled using an auger until the top of the kaolin is reached and then a core barrel is used to drill through the kaolin bed. If

Table 10. Purpose of kaolin drilling program

Drill spacing (M)	Purpose
120	Property exploration
60	Proving the kaolin reserve
30	Mine planning and development
15	Quality control, mine inventory volume, and tonnage measurement

the wider spaced drilling indicates that the size and quality of the deposit is sufficient to warrant mining, then close spaced drilling is done (Table 10). The information from the auger and core drilling provides the thickness and type of overburden and the quality and thickness of the kaolin, so that the stripping ratio can be determined. The stripping ratio is the overburden thickness over the kaolin thickness. Stripping ratios determine whether or not mining the deposit is economical. The lower the ratio, the lower the cost of mining. The drilling program also is used to evaluate potential groundwater problems if the water table is high or if there is artesian water in sand bodies immediately under the kaolin bed. In kaolin or bentonite deposits of hydrothermal origin, the drill hole locations are based on the topography and shape and size of the alteration zone. Fig. 23 shows the configuration of a hydrothermal kaolin deposit in the Cornwall area of southwestern England, which requires a special drill hole pattern to delineate the deposit.

Flat lying bentonite deposits are also drilled using a grid pattern, but if the deposits are steeply dipping or structurally deformed, then special drilling patterns are used. Auger drilling bentonite deposits are much more commonly used than core drilling. The same is true for drilling palygorskite and sepiolite deposits. Normally, the palygorskite and sepiolite deposits are flat lying as are ball clay deposits. They are sedimentary clays that were deposited in lacustrine, swamp, or tidal flat environments, and are located in areas that have not been deformed by mountain building, faulting, and folding.

1. KAOLIN MINING AND PROCESSING

Once a mine plan has been determined after the drilling is completed, the land is cleared and the removal of overburden begins. Open pit methods of mining are used in the major kaolin deposits around the world. A variety of stripping methods is used including hydraulic backhoes and shovels which load directly into large off-road trucks; pan-type self-loading scrapers that are sometimes pushed by dozers if the overburden dirt is wet or relatively dense; large draglines which dump the spoil into the previously mined out panels; and bucket wheel excavators loading the spoil on conveyor belts. This latter method is sometimes used in Europe.

After the overburden is removed, the kaolin is mined (Murray, 1963) using much the same methods that are used to remove the overburden. However, the mining must be done with much more care to assure quality and in order to recover as much ore grade clay as possible. In a single mine, there may be significant quality differences so selective mining and segregation are often necessary. The usual practice is to mine a particular quality of kaolin and transport it to a stockpile, which is made up of a similar quality. Several stockpiles are built based on brightness, color, viscosity, and grit percentage.

Once the kaolin is in a particular stockpile, the wet processing (Murray, 1980; Pruett, 2000; Kogel et al., 2002) is initiated. A generalized flow sheet for the wet process is shown in Fig. 45. Kaolin from a graded stockpile is hauled to a blunger where it is mixed with water and a small percentage of a chemical dispersant. The percent solids in the blunger ranges from 40% to 60% although the lower solids is much more common. Kaolin from a single stockpile can be blunged and kaolin from multiple stockpiles can be blended and blunged to achieve a particular quality. The blunger, which can be stationary or portable, is fed using front-end loaders or in some cases, with a small dragline shovel. The blunger is a high speed, high

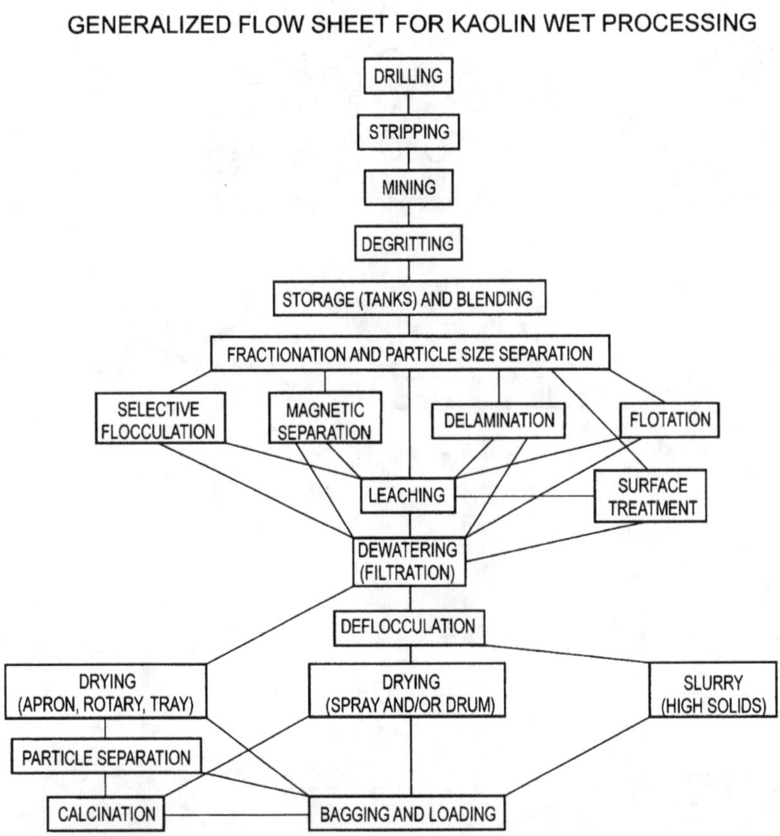

Fig. 45. Wet process flow sheet.

horsepower mixer which breaks up the kaolin lumps into discrete individual particles. A dispersant (Murray, 1984) is necessary in order to keep the discrete particles separated from each other because otherwise the particles would flocculate. Fig. 46 is a diagram showing flocced particles and dispersed particles. Fig. 47 is a diagram showing the charges on the crystals of kaolinite. Because of the positive and negative charges, the kaolinite particles are attracted and form large aggregates or flocs. The addition of a soluble dispersant which ionizes to produce cations that are attracted to the negative charges on the clay particle so that each kaolinite plate or stack has a similar charge and thus they repel each other. The most commonly used chemical dispersants are sodium silicate, sodium hexametaphosphate, tetrasodium pyrophosphate, and sodium polyacrylate. The amount of dispersant added is quite small, of the order of 4–12 lb/ton of kaolin, which is 0.2–0.6% based on the dry weight of the kaolin.

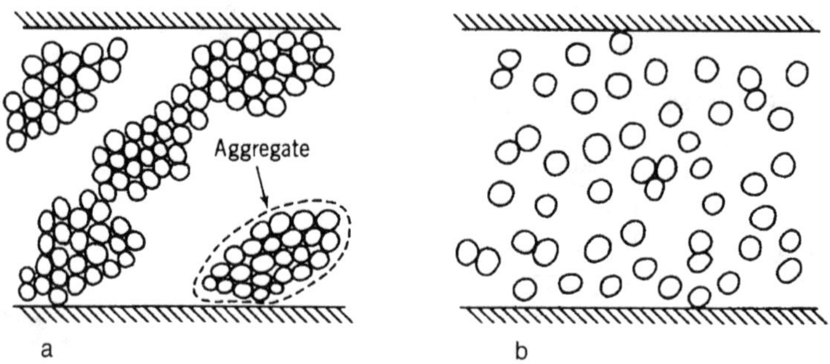

Fig. 46. (a) Flocced and (b) Dispersed particles.

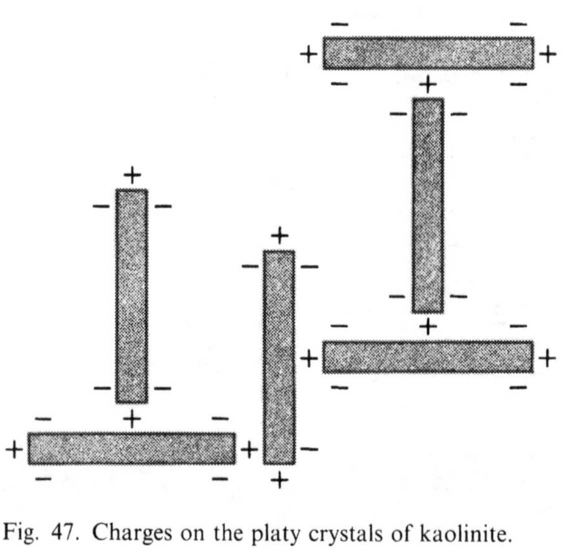

Fig. 47. Charges on the platy crystals of kaolinite.

Once the kaolin is blunged and dispersed into a slurry, the next step in the process is to remove the grit. Grit is defined as particles coarser than 325 mesh or 44 μm. The grit in kaolin is usually comprised of quartz sand, mica, and a suite of heavy minerals (Murray, 1976). A common method for removing grit is to pass the slurry through drag boxes, which are known as sandboxes. A residence time of about 30 min is adequate to allow the coarse grit particles to settle to the bottom of the drag box. These coarse settled impurities are then removed by drag slats and disposed of in waste impoundments. Mica, which is flake shaped, does not settle as rapidly as the quartz and heavy minerals so the slurry goes from the drag box to a vibratory screen which removes the coarse mica and other floating debris that may be present. Hydrocyclones are sometimes used instead of drag boxes, particularly if the grit percentage is higher than about 15%. Hydroseparators are also used to remove grit.

After degritting, the slurry is pumped to large mine holding tanks, which when filled and checked for quality, is then pumped through a pipeline to terminal tanks at the processing plant. The mine holding tanks are also used to blend kaolins in order to meet viscosity and brightness specifications. The longest pipeline in Georgia is about 35 miles (56 km) in length and the longest in Brazil is about 100 miles (160 km) in length. Further blending, if necessary to meet quality specifications, can be accomplished in the terminal tanks at the processing plant.

The next step in the wet process (Fig. 45) is to fractionate the kaolin into coarse and fine fractions. This is accomplished by continuous bowl-type centrifuges, hydroseparators, or hydrocyclones. After fractionation to a particular particle size, the fine fraction and the coarse fraction of the kaolin are pumped to holding tanks. The coarse fraction may be delaminated (which will be described later in this chapter) or is filtered and dried to produce filler clays. The fine fraction can then be passed through a high intensity magnetic separator which removes discrete iron and titanium minerals. Other processes used to remove the iron containing titanium minerals, usually anatase, are selective flocculation and flotation. These processes will be described later in this chapter. The fine fraction slurry can go through one of the above processes before going to the floc and leach step or it can go directly to floc and leach depending on the brightness of the grade to be produced. The floc and leach step is to acidify and floc the slurry at a pH between 2.5 and 3, which solubilizes some of the iron compounds which stain the kaolin. Alum is sometimes used in combination with sulfuric acid to give a tighter floc. At essentially the same time, a strong reducing agent, sodium hydrosulfite, is added to the slurry to reduce ferric iron to ferrous iron, which then combines with the sulfate radical to form a soluble iron sulfate, $FeSO_4$. The iron sulfate is removed in the filtration step, which is the next step in the process. Quality control determines the quantity of acid, alum, and hydrosulfate that is needed to give the best brightness result.

After the floc and leach process, the flocced slurry is pumped to filters to remove water and the soluble iron sulfate. Usually water spray bars are used to wash the filter cake to remove more of the iron sulfate. Commonly, the percent solids after the floc and leach is around 25%. Large rotary vacuum filters or plate and frame pressure filters are used to dewater the kaolin, raising the percent solids to 60–65%. After filtration, the filter cake is redispersed and pumped to a spray drier where it is dried for bulk or bag shipments or the percent solids is increased to 70% by adding dry spray dried clay or by large evaporators which is the slurry solids necessary for most tank car or tank truck shipments. The filter cake can be extruded and dried to make what is termed an acid kaolin product.

The coarse fraction from the centrifuges is used either to make coarse filler clays or as feed to produce delaminated kaolins (Fig. 48). The coarse thick vermicular stacks and books of kaolin are pumped to delaminators which shears the plates making up the stack or book into large diameter thin plates (Kraft et al., 1972). These large diameter thin plates have what is termed a high aspect ratio which is a ratio of the diameter to the thickness of the plate. The stacks and books have a prominent cleavage, which is parallel to the (001) basal plane. The coarse particles are cleaved by placing them in a baffled vessel filled with media in which impellers strongly agitate the slurry. The spherical media which can be used is well-rounded sand, alumina proppants, and/or glass, plastic, zirconia, or alumina beads. This vigorous agitation of the media and the coarse kaolin cause the kaolin to shear upon collision between the media beads to produce a coarse delaminated plate with a high aspect ratio (Fig. 49).

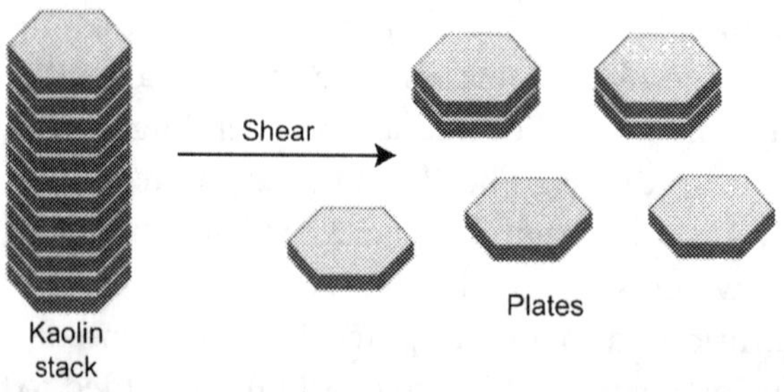

Fig. 48. Delamination.

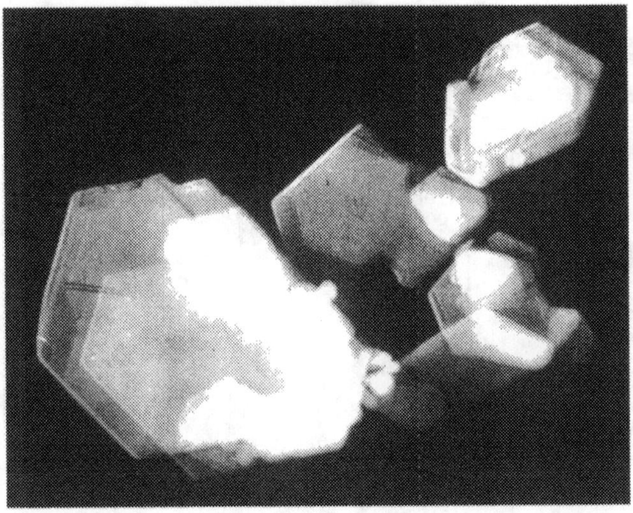

Fig. 49. Electron micrograph of delaminated kaolin plates.

The magnetic separation process involves the use of powerful magnets with field strengths ranging from 2 to 6 T. The range from 2 to 6 T is achieved by using liquid helium cooled superconducting coils which results in considerable savings in electric power. The kaolin slurry is pumped through a highly compressed fine stainless steel wool matrix, which when energized, separates the magnetic minerals and allows the non-magnetic kaolinite to pass through the matrix. The magnetic field is periodically switched off so that the accumulated magnetic particles can be rinsed with water, thus cleaning the steel wool matrix. Fig. 50 is a diagrammatic representation of a 2 T magnet. The magnetic minerals that are removed are dominantly hematite and yellowish iron enriched anatase along with some ilmenite, magnetite, and biotite. The magnetic separation process was described by Iannicelli (1976) who was one of the first to advocate the use of magnetic separation in order to brighten kaolin clays. The development of high intensity wet magnetic separation for use in the kaolin industry has resulted in a huge increase in kaolin reserves which can be used commercially (Murray, 2000).

The froth flotation process used to remove dark iron stained anatase which discolored the kaolin was initially developed by Greene and Duke (1962). They used a calcium carbonate carrier which was termed a "piggy back" process. Since then, the flotation process has been improved so that now it has evolved into a standard method in processing Georgia kaolins to make high brightness products of 90% or higher. The dark iron stained anatase is selectively coated with a reagent which causes it to

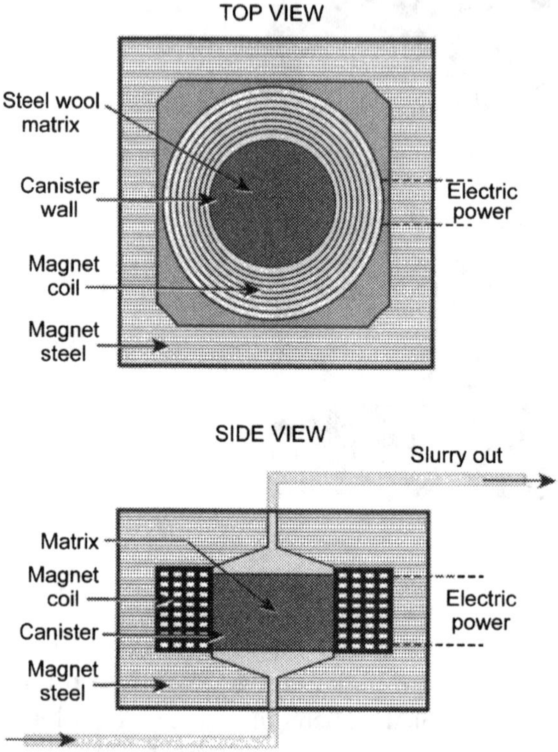

Fig. 50. Diagrammatic scheme of 2 T magnet.

adhere to air bubbles sprayed into the slurry. The air bubble froth which contains the stained anatase rises to the surface of the float cell and is skimmed off and discarded. Denver-type conditioners and float cells are the most commonly used equipment. Recently, vertical column flotation cells have been used which improves the separation of fine particles and also increases product recovery. Most of the Georgia kaolins contain up to 2.5% T_iO_2 and by using the flotation process, the percentage can be reduced to as low as 0.3.

Selective flocculation is another process that can be used to reduce the T_iO_2 percentage. The process was introduced in the late 1960s by Bundy and Berberich (1969) to produce high brightness products of 90% or higher. Since its initial development, the selective flocculation process has been continually improved and is now a process which is used extensively to produce high brightness products (Shi, 1986, 1996; Pruett, 2000). This process is the reverse of flotation in that the dark iron stained anatase is selectively flocculated so that it settles in a hydroseparator while the kaolin remains suspended in a dispersed condition. The flocculated anatase is discarded into waste impoundments.

Another special process used to produce value-added products is calcination, which was introduced in the early 1950s. The kaolinite is processed to remove impurities and a fine particle size gray kaolin is a preferred feed (Fanselow and Jacobs, 1971). The fine gray kaolin is spray dried, pulverized, and then fed to either rotary or large hearth calciners and heated to as high as 1300°C. The highest temperature of 1300°C is used to produce granules for use in making refractory shapes and bricks. Most of the pigment grade of calcined kaolin is heated to a temperature between 1000 and 1050°C. Fig. 51 shows the temperature at which the kaolin is dehydroxylated to form metakaolin which is then transformed into mullite (Fig. 52). The metakaolin is an amorphous mixture of alumina and silica that is used in several applications which are described in Chapter 5. The phase change at 980°C transforms the amorphous metakaolin into mullite (Al_2SiO_5). This causes a significant increase in brightness and opacity which is also discussed in Chapter 5.

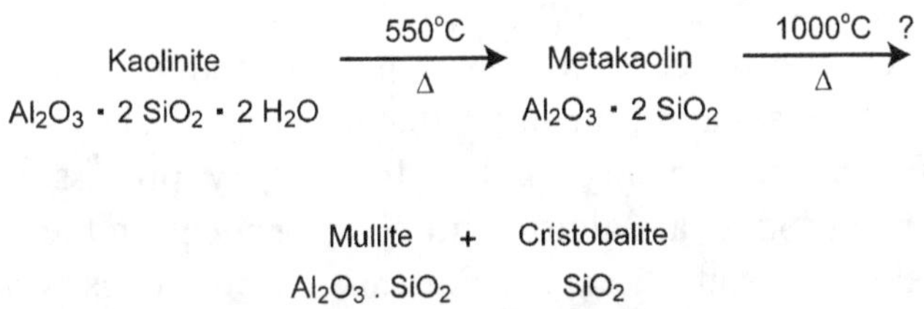

Fig. 51. Calcination temperature.

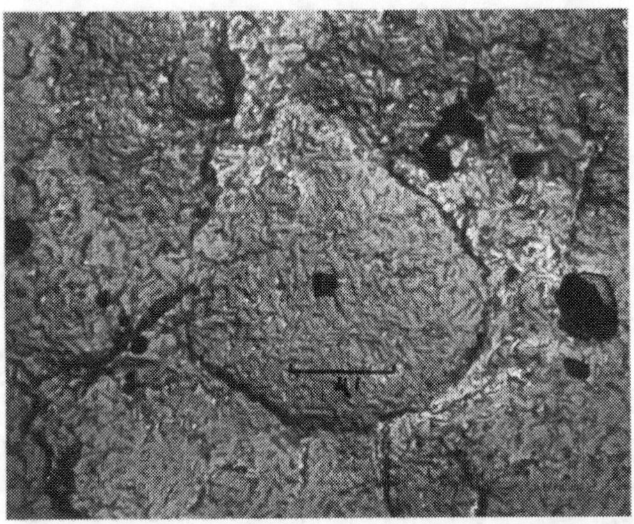

Fig. 52. Calcined kaolin surface.

The hardness of the calcined kaolin is about 6.5 on the Mohs scale, which is considerably harder than the 1.5–2 hardness of hydrous kaolin. An 85% brightness feed to the calciner will produce a product with a brightness of 91–93%.

Special processes are used to modify the surface properties of kaolinite in order to improve the functionality and dispersion of the product (Grim, 1962; Nahin, 1966; Libby et al., 1967; Bundy et al., 1983; Iannicelli, 1991). The hydrophilic surface of kaolinite can be chemically treated to make them hydrophobic or organophilic. These surface modified kaolins can then be used as a functional pigment and/or extender in systems where the natural hydrophilic kaolin cannot be used. The uses of these surface modified kaolins are discussed in Chapter 5.

2. DRY PROCESS

Some kaolin is dry processed (Murray, 1982), which is simpler and less costly than the wet process. Lower cost and lower quality products can be used, for example, in fiberglass and cement production. Fig. 53 shows a typical flow sheet for dry processing kaolin. In the dry process, the properties of the kaolin product are almost entirely dependent on the crude clay quality as delivered from the mine. For this reason, deposits

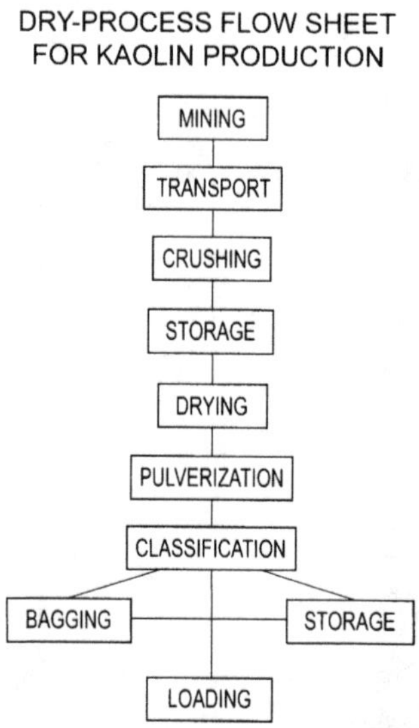

Fig. 53. Dry process flow sheet.

must be selected that have the brightness, grit percentage, and particle size distribution that can be dry processed to make a particular product. The upper limit of grit percentage that can be handled in the dry process is usually about 7%.

The stripping and mining are similar to that described previously for the wet process. The mined kaolin is transported to the processing plant where it is crushed or shredded and placed in large storage sheds partitioned into bays in which a particular quality is stored. The size of the crushed or shredded kaolin particles is egg size or smaller. These egg-sized lumps of kaolin are fed into a rotary drier which reduces the moisture to 6% or less. The dried kaolin is pulverized in roller or hammer mills or some other disintegrating device. Heated air can be used in this step to further dry the pulverized product if necessary. The pulverized kaolin is commonly air classified to remove grit size particles. Also, fine particle size products can be produced using an air classification system. The product is then shipped in bulk or in bags to the customer.

3. HALLOYSITE MINING AND PROCESSING

As mentioned previously in Chapter 3, a currently operating halloysite mine is located (Fig. 37) on the North Island of New Zealand (Murray et al., 1977). The halloysite in New Zealand is hydrothermally altered from rhyolite on which surficial weathering has been superimposed. The drilling of the halloysite deposit was done with a core drill with an initial grid pattern of 30 m. Subsequent drilling is done on a 15 m spacing particularly to determine the quality and useable thickness. The deeper altered material is not as high quality as that in the upper portion of the deposit which was further altered by surficial weathering. The halloysite is mined with a hydraulic shovel which loads the clay into trucks, which transports it to a stockpile at the plant. The halloysite is blunged, dispersed, and degritted similar to the methods used by the kaolin industry in Georgia. The degritted slurry is further processed using a sand grinder similar to that described to delaminate the kaolin. This is done to fully separate and disperse the halloysite so that a 2 μm particle size product can be produced. After the sand grinder, the slurry is centrifuged to separate and recover a 2 μm function which is then leached, filtered, and dried. The coarse fraction is used for local ceramic manufacturing and as filler in paper. The fine fraction is used as an additive in making high quality dinnerware (Harvey, 1996).

4. BALL CLAY MINING AND PROCESSING

Ball clay deposits are generally smaller in areal extent than the sedimentary kaolins. The exploration, stripping, and mining are similar. The ball clay is transported by truck from the mine to large covered storage sheds with bins to separate the crude ball clay based on ceramic quality considerations such as color, plasticity, particle size, green, dry, and fired strength and fired color. A dry process similar to the kaolin dry process and a partial wet process (Fig. 54) are used to process the ball clay. The dry process is very similar to the dry process used in the kaolin industry. Considerable shipments of ball clay are now in slurry form in tank cars and tank trucks. The crude ball clay is shredded or crushed and blunged in the same type of blunger used in the kaolin wet process. The clay, water, and chemical dispersant are mixed and blunged and then screened to remove grit and other debris. The dispersed ball clay slurry is then pumped into large holding tanks from which the tank cars and tank trucks are loaded. Blending to meet certain required specifications can be done in these holding tanks. The percent solids of the finished slurry ranges between 60% and 65%, dependent on the viscosity. The slurried ball clay is of better quality compared to the dry process product because it is relatively free from grit and can be blended more easily to give the customer a more uniform and higher-quality product. Some of the dry process ball clay is shredded and dried and shipped to the customer in lump form without further processing. The pulverized clay can be air separated to remove coarse grit. The shredded and dried ball clay is shipped at about 12% moisture in bulk or bags to the customer. The pulverized ball clay has a moisture content of 3% or less.

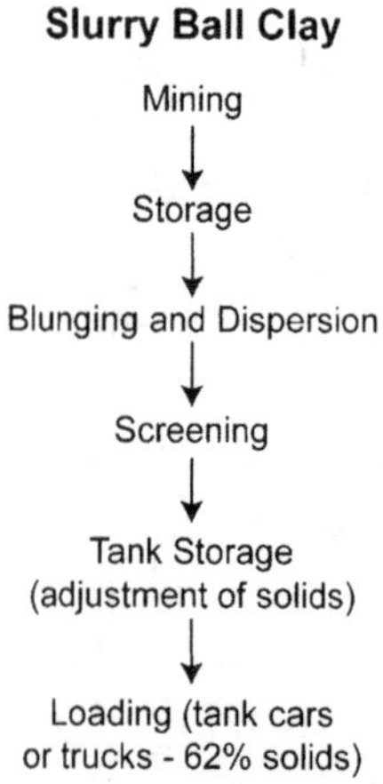

Fig. 54. Slurry process for ball clay.

5. BENTONITE EXPLORATION, MINING, AND PROCESSING

Most bentonite exploration drilling is done using auger drills. The spacing between drill holes is determined based on the topography, dip of the beds, and overburden thickness. Once the deposit is drilled and tested and is found suitable for mining, the first cut is stripped using backhoes and trucks, motorized scrapers, bulldozers, or draglines. Once the stripping is completed, the crude bentonite is either mined and placed in a stockpile at the mine area or is hauled to the processing plant and stockpiled. In many instances, the physical properties of the bentonite are improved by aging and oxidation. Examples are increases in viscosity for drilling muds and green strength for bonding foundry sands and iron ore pellets. Some of the stockpiles weather and oxidize for as long as a year. Often the stockpile at the mine is turned over by plowing to expose unweathered clay. As soon as the bentonite is weathered and has the quality needed, the dry process (Fahn, 1965) begins. Normally, the bentonite is loaded into a hopper which feeds a shredder or crusher to reduce the size of the crude lumps to about 1 in. Some crude bentonite is pugged at this stage in order to provide a more uniform feed to the drier. Rotary driers are normally used but fluid bed driers may be used in order to obtain more uniform drying so that the lumps do not get overheated. Overdrying results in a loss of some physical properties such as viscosity and plasticity. After drying, the dried lumps are passed through a roll crusher to produce granules, which are then screened using decked screens to produce granules of a particular size such as 15/30 which means that the granules are coarser than 15 mesh and finer than 30 mesh. Finer granular products are also produced. The granules are normally about 6–10% moisture. They are then bagged or loaded in bulk in hopper cars.

The dried lumps can bypass the roll crusher and be fed into a roller mill or impact mill to produce a pulverized product for bagging or shipping in bulk. Sometimes heat is used in the pulverization process to further dry the products to a specified moisture content. Also, soda ash or sodium polyacrylate can be added to enhance the swelling properties and increase the viscosity.

A small percentage of bentonite can be water washed for use in special applications such as producing organoclays. The wet process involves blunging the bentonite at low solids, screening or centrifuging to remove the grit, centrifuging to produce a very fine particle size and treatment with a specific chemical to make an organoclay and then either flash drying or drum drying. The temperature of drying must be controlled in order that the organic compound such as a quaternary amine is not destroyed. Because the montmorillonite contains water between the 2-to-1 structural layers, it is fairly easy to introduce polar chemical compounds to replace the water and make the particles organophilic or hydrophobic for special applications which will be discussed in Chapter 6.

Bleaching earths are calcium montmorillonites which have been processed to purify, bleach, and clarify certain liquids. The fine particle size, the high charge on the layer and the high surface area after processing make certain calcium bentonites excellent acid-activated clays. In Europe, the calcium bentonites are known as fuller's earth, whereas in the United States, the term is used to describe any natural material able to decolorize and bleach oils and absorb water.

Acid activations are usually produced wet (Fig. 55), but in recent years, some bleaching earths are produced using a dry process. The dry process involves crushing, drying, pulverization, acid treatment, and packaging. The wet process involves blunging the calcium montmorillonite at low solids normally 20% or less, screened or hydrocycloned to remove the grit, heating the slurry to a specific temperature normally around the boiling point, adding sulfuric or hydrochloric acid, dewatering in a filter press, and dried and formed into a powder or granules. If the product is a fine powder, flash drying is commonly used. Sulfuric acid is preferred because it is less expensive and is not as harsh as hydrochloric acid. The acid treatment increases the surface area and pore volume. The digestion time after the acid addition must be controlled in order not to destroy the crystalline structure of the calcium montmorillonite. Crossley (2001) estimated the bleaching clay market was about

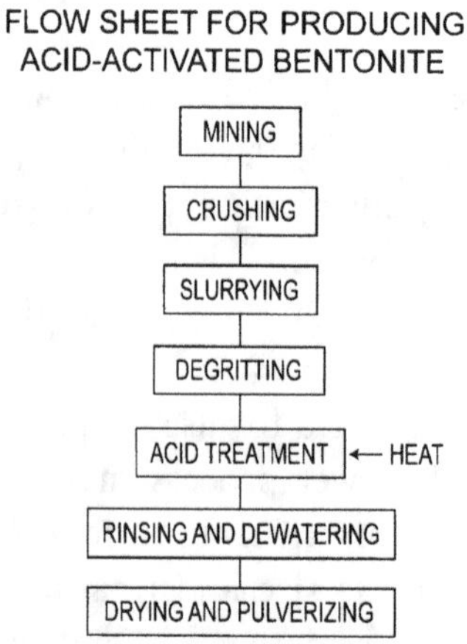

Fig. 55. Acid activation process.

860,000 tons annually. The applications of bleaching clays will be discussed in Chapter 6.

6. PALYGORSKITE AND SEPIOLITE EXPLORATION, MINING, AND PROCESSING

Both palygorskite and sepiolite are auger and core drilled in the exploration phase to determine the size of the deposit, the thickness of the overburden, and the quality. If the deposit is thought to be a useable deposit, then it is drilled to determine the lateral and vertical variability. After the deposit is determined to be mineable, then the deposit is stripped with conventional open pit stripping methods. Once a substantial area of clay is stripped, then closely spaced core drilling is done for mine quality control. The cores are taken to the laboratory and tested to determine the most suitable applications for the finished products. Based on these tests, the palygorskite or sepiolite is blocked out and selectively mined, stockpiled, and processed through the plant (Oulton, 1963).

The clay is mined with power shovels and hoes and loaded into trucks for transportation to plant crude clay storage stockpiles. The crude clay contains from 40% to 50% volatile matter which is principally free and has combined water. Fig. 56 shows a typical processing flow sheet. The crude clay is first crushed and either goes directly to the driers or is extruded. Extrusion brings about a marked modification and imparts properties which are highly desirable for certain applications. Sometimes

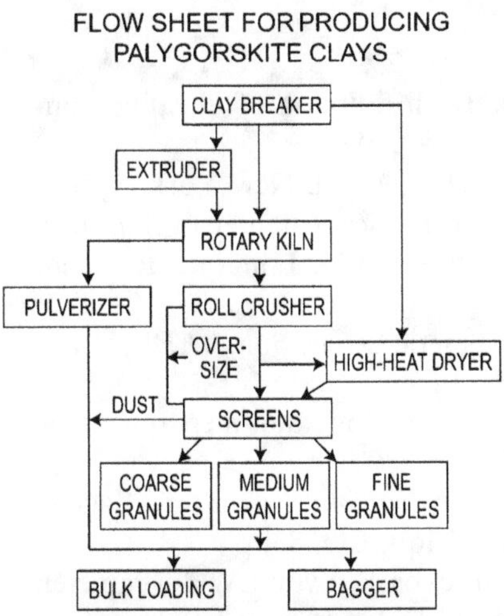

Fig. 56. Palygorskite flow diagram.

MgO is added at the extruder which improves the viscosity. The extrusion separates the bundles of the palygorskite into separate particles. Palygorskite and sepiolite dried at moderate temperatures retains colloidal properties, while higher temperature drying develops useful absorptive and other properties, so two driers are used, which are regulated at a relatively low and a high temperature usually about 200 and 600°C, respectively. After drying, the clay goes to roll crushers and then to screens where granular products are separated. After drying, the lumps can go directly to pulverizers to produce extra fine products. The granular products are coarser than 100 mesh and the most common granular grades are 15/30, 30/60, and 60/90. Medium and fine grades range from 100 to 325 mesh. Still finer grades are pulverized to a particle size of 95% finer than 10 μm.

Palygorskite and sepiolite may also be used as a natural bleaching earth, which does not require acid activation. It is used as a purifying and decolorizing agent in petroleum refining by percolation processes.

REFERENCES

Bundy, W.M. and Berberich (1969) *Kaolin Treatment*. US Patent 3477809.

Bundy, W.M., et al. (1983) Chemically induced kaolin floc structures for improved paper coating. *Proceedings of the Technical Association of the Pulp and Paper Industry*. Ind. Coating Conf., San Francisco, CA, pp. 175–187.

Crossley, P. (2001) Clear opportunities for bleaching and clarifying clays. *Ind. Miner.*, 402, 69–75.

Fahn, R. (1965) The mining and preparation of bentonite. *Int. Ceram.*, 12, 119–122.

Fanselow, H.R. and Jacobs (1971) US Patent 3586523.

Greene, E.W. and Duke, J.B. (1962) Selective froth flotation of ultrafine minerals or slimes. *Trans. Soc. Min. Eng. AIME*, 223, 389–395.

Grim, R.E. (1962) *Applied Clay Mineralogy*. McGraw-Hill, New York, 375pp.

Harvey, C.C. (1996) Halloysite for high quality ceramics. Chapter in *Industrial Clays*, 2nd Edition. Industrial Minerals Info. Ltd., London, England, pp. 71–73.

Iannicelli, J. (1976) New developments in magnetic separation. *IEEE Trans. Magn. Mag.*, 12, 436–443.

Iannicelli, J. (1991) Polymer reinforcement with amino and mercaptosilane grafted kaolin. *Miner. Tell. Process*, 8, 135–138.

Kogel, J.E., et al. (2002) *The Georgia Kaolins—Geology and Utilization*. Society for Mining, Metallurgy and Exploration, Littleton, CO, 83pp.

Kraft, K., et al. (1972) Die Unterschiedliche Bedruckbarkeit von LWC—Papieren Beim Einsatz Herkomlicher Strichkaoline bzw. Delaminierter Kaoline. *Sunderdruck av dem Wochenblatt für Dapier Fabrication*, 100, 131–136.

Libby, P.W., et al. (1967) Elastomer reinforcement with amino silane grafted kaolin (Abstract). American Chemical Society, 154 (Division of Rubber Chemistry).

Murray, H.H. (1963) Mining and processing industrial kaolins. *Georgia Miner. Newslett.*, XVI(1–2), 12–19.

Murray, H.H. (1976) The Georgia sedimentary kaolins. *Proceedings of the 7th International Kaolin Symposium*, Shomodu, S. ed., University of Tokyo, Tokyo, pp. 114–125.

Murray, H.H. (1980) Major kaolin processing developments. *Int. J. Miner. Process.*, 7, 263–274.

Murray, H.H. (1982) Dry processing of clay and kaolin. *Inter. Ceram.*, 31(2), 108–110.

Murray, H.H. (1984) Kaolins. Chapter in *Paper Coating Pigments*. Hagemeyer, R.W., ed. Tappi Press, Atlanta, GA, pp. 95–141.

Murray, H.H. (2000) Traditional and new application for kaolin, smectite, and palygorskite: a general overview. *Appl. Clay Sci.*, 17, 207–221.

Murray, H.H., et al. (1977) Mineralogy and geology of the Maungaparerua halloysite deposit in New Zealand. *Clay. Clay Miner.*, 25, 1–5.

Nahin, P.G. (1966) US Patent 3248314.

Oulton, T.D. (1963) Commercial production and uses of attapulgite clay products. *Georgia Miner. Newslett.*, XVI(1–2), 26–28.

Pruett, R.J. (2000) Georgia kaolin: development of a leading industrial mineral. *Min. Eng.*, 52, 21–27.

Shi, J. (1986) *Method of Beneficating Kaolin Using Ammonium Salt*. US Patent 4604369.

Shi, J. (1996) *Processing for Removing Impurities from Kaolin*. US Patent 552986.

Chapter 5

KAOLIN APPLICATIONS

Kaolin is one of the more important industrial clay minerals. Kaolin is comprised predominantly of the mineral kaolinite, a hydrated aluminum silicate. As noted in Chapter 2, other kaolin minerals are dickite, nacrite, and halloysite. Dickite and nacrite are rather rare and usually are found mixed with kaolinite in deposits of hydrothermal origin. Relatively pure halloysite deposits are rare and as was pointed out in Chapter 4, one of the only commercial halloysite deposits now operating is located on the North Island of New Zealand. The Dragon halloysite mine in Utah was operated for many years and then was abandoned. However, it is being reopened as additional reserves have been located, so it may become another source of commercial halloysite (Wilson, 2004).

Kaolinite, which is the dominant mineral in kaolin deposits, is a common clay mineral, but the relatively pure and commercially useable deposits are few in number. Kaolinite has physical and chemical properties which make it useful in a great number of applications.

In contrast to smectites and palygorskite and sepiolite, kaolinite is less reactive when incorporated into most industrial formulations which accounts for many of its more important applications. Such characteristics as low surface charge, relatively low surface area, white color, low ion exchange, and particle shape make it a prime pigment and extender in paper coating and paints and other specialty applications. An example of the difference in the clay mineral types is in their viscosity in water. Relatively pure kaolinite has a low viscosity at very high solids content up to 70% or slightly higher. Sodium montmorillonite, in contrast, has a very high viscosity at 5% solids because of its high surface charge, surface area, exchange capacity, and very fine particle size. Palygorskite and sepiolite have a high viscosity because of their elongate particle shape. Again, it is the fundamental structure and composition that controls the resultant physical and chemical properties which are important in determining their many industrial applications. The more important properties are listed in Table 11.

Table 11. Important properties of kaolin

1	White or near-white in color
2	Chemically inert over a wide pH range (4–9)
3	Fine in particle size
4	Soft and non-abrasive
5	Platy with the plate surface dimensions relatively large compared to the thickness
6	Hydrophilic and disperses readily in water
7	Because of its shape, it has good covering and hiding power when used as a pigment or extender in coatings
8	Plastic, refractory and fires to a white or near-white color
9	Low conductivity of both heat and electricity
10	A very low charge on the lattice
11	A low surface area as compared with other clay minerals
12	Some kaolins have a low viscosity and flow readily at 70% solids
13	Relatively low in cost

Table 12. Representative physical constants of kaolinite

Specific gravity	2.62
Index of refraction	1.57
Hardness (Mohs' scale)	1.5–2.0
Fusion temperature (°C)	1850
Einlehner abrasion number	4–10
Dry brightness at 457 nm (%)	75–93
Crystal system	Triclinic

All the properties listed in Table 11 contribute to the many applications of kaolin. Table 12 gives the representative physical constants of kaolinite. It is estimated that worldwide, some 40,000,000 tons annually are mined and processed. Table 13 shows typical chemical analyses of a Georgia soft and hard kaolin, an English primary kaolin, a Brazil soft and hard kaolin, and a theoretical kaolinite.

1. PAPER

One of the most important applications of kaolin is coating and filling paper. As a filler, the kaolin is mixed with the cellulose fibers in wood pulp and as a coating, the kaolin is mixed with water, adhesives, and various additives and coated onto the surface of the paper. The coating makes the paper sheet smoother, brighter, glossier, more opaque, and most importantly, improves the printability (Bundy, 1993). Paper that is not coated is made up of cellulose fibers interwoven in a random and open configuration. Uncoated paper does not meet the stringent

Table 13. Typical chemical analyses of some kaolins (wt.%)

Component	Cretaceous middle Georgia kaolin	Capim soft kaolin	Tertiary East Georgia kaolin	Jari hard kaolin	Theoretical kaolin
SiO_2	45.30	46.56	44.00	44.45	46.3
Al_2O_3	38.38	38.03	39.5	37.37	39.8
Fe_2O_3	0.30	0.59	1.13	1.93	
TiO_2	1.44	0.78	2.43	1.39	
MgO	0.25	0.01	0.03	0.02	
CaO	0.05	0.01	0.03	0.01	
Na_2O	0.27	0.03	0.08	0.01	
K_2O	0.04	0.02	0.06	0.12	
Ignition loss	13.97	13.8	13.9	14.45	13.9

requirements for high quality printing and particularly multicolor printing. The fine particle size and platy shape of kaolinite are ideal for imparting a smooth, dense surface that is uniformly porous. This gives the paper a more uniform ink receptivity.

The hydrophilic nature of kaolinite makes it easily dispersable in aqueous systems. Coating formulations consist of pigment, binder, water, and small amounts of other additives. This formulation, called a coating color, is metered onto the paper surface with a trailing blade coater or other types of coaters. The shear values at the coating blade interface are extremely high because the paper travels at speeds as high as 1500 m/min. The coating color rheology should be Newtonian or thixotropic (Fig. 57) so that the coating spreads readily on the paper. If the clay is dilatant then pinheads develop which cause streaks on the coated paper.

Optical properties of coatings are brightness, gloss, and opacity (hiding power). Brightness of the paper is largely a function of the brightness of the grade of kaolin used. Gloss increases with decrease in particle size. Opacity is controlled by light scatter, which is dependent on the difference in the refractive index of the kaolinite and of air-filled voids (Fig. 58). Particle size distribution and the amount of fines of the order of 0.25 μm have a large influence on the opacity.

Relatively fine particle size kaolin products of the order of 80% less than 2 μm or finer are the grades that are used in paper coatings. Delaminated kaolins are favored in lightweight coatings (LWC). The relatively large diameter of delaminated particles impart a shingle-like structure to coatings which gives good ink holdout and smoothness. The LWC have reduced the weight of the paper so that postal rates are lower for many

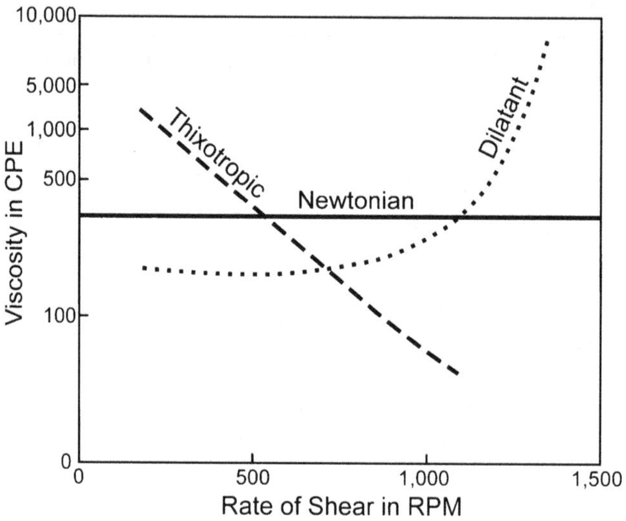

Fig. 57. Rheology—dilatant, Newtonian, and thixotropic.

$$\text{Reflection coefficient, } R = \frac{(n_1 - n_0)^2}{(n_1 + n_0)^2}$$

n_1, refractive index of pigment
n_0, refractive index of adhesive or air

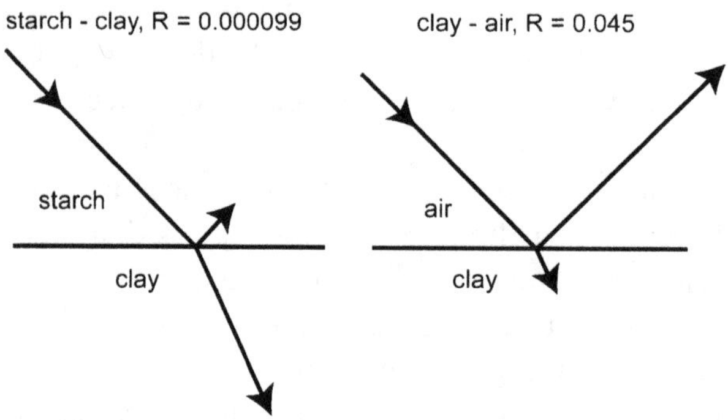

Fig. 58. Opacity.

magazines such as the weekly news magazines. Fig. 59 is an electron micrograph of a delaminated kaolin-coated paper and Table 14 shows many of the coating grades of kaolin and their particle size and brightness.

Another development in paper-coating clay is the production of engineered or tailored products (Murray and Kogel, 2005). These products are engineered to enhance specific properties such as opacity, gloss, brightness, ink holdout, whiteness, and print quality. This can be

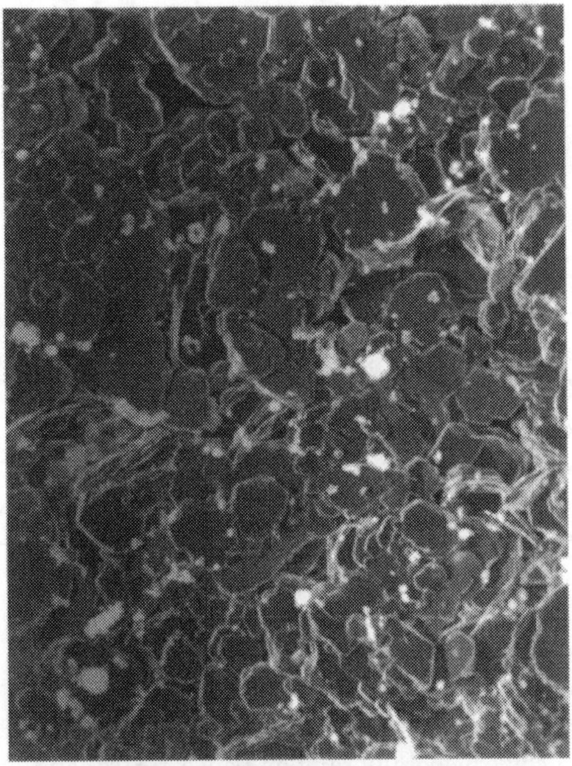

Fig. 59. SEM paper coated with delaminated kaolin.

Table 14. Particle size and brightness of some coating kaolin clays

	Particle size	GE brightness
Regular coating clays		
No. 3	72% < 2 μm	8.5–86.5
No. 2	80–82% < 2 μm	85.5–87
No. 1	90–92% < 2 μm	87–88.0
Fine No. 1	95% < 2 μm	86–87.5
Delaminated coating clays		
Regular	80% < 2 μm	88.0–90.0
Fine	95% < 2 μm	87.0–88.0
High brightness coating clays		
No. 2	80% < 2 μm	89.0–91.0
No. 1	92% < 2 μm	89.0–91.0
Fine No. 1	95% < 2 μm	89.0–91.0
Special engineered clays	80–95% < 2 μm	90.0–93.0
Calcined kaolins	88–95% < 2 μm	92.0–95.0

accomplished by processing the kaolin to a specific particle size distribution, brightness, increased aspect ratio, and control of the percentage of both the coarse and fine particle sizes. The closer that a particle size distribution is between 2 and 0.5 μm, the better the optical properties (Bundy, 1967).

Rheology (Murray, 1975) is a very important property to control for use in paper-coating formulations. Both low shear and high shear viscosity are important. Stringent viscosity specifications are set for coating clays. Factors which determine viscosity are particle size and shape, surface area and charge, mineralogical impurities, and chemical impurities (Lagaly, 1989; Bundy and Ishley, 1991). Morphology is an important factor in the viscosity of kaolin suspensions (Yuan and Murray, 1997). The presence of montmorillonite, mica, or halloysite is detrimental to good viscosity (Pickering and Murray, 1994).

A kaolin-based pigment having high surface area has been developed for ink jet matte-coating applications (Malla and Devisetti, 2005). Its unique morphology allows high solids dispersion with either anionic or cationic dispersants, yet has better viscosity than silica-based pigment slurries.

Kaolins used as fillers in paper are relatively coarse, ranging between 40% and 60% less than 2 μm. The brightness of the filler clays is normally less bright than coating clays, generally ranging between 80% and 85%. The coarse kaolin particles are mixed with the paper pulp or fed from headboxes onto the wet pulp, which is layered onto a wire mesh belt. The kaolin particles are trapped in the interstices of the cellulose fibers. The clay filler improves the brightness, opacity, smoothness, ink receptivity, and printability. A perfect filler, if available, would have these characteristics (Willets, 1958) (Table 15).

Kaolin, of course, is not a perfect filler, but meets several of the criteria listed in Table 15. It is used in white papers such as newsprint, printing grades, and uncoated book paper. Cost reduction is an important factor as the filler is much less expensive than the pulp it replaces. Table 16 shows filler grades of kaolin.

Rheology is relatively unimportant in paper filling except in the dispersion and pumping of the kaolin slurry. Up until about 1980, kaolin was the dominant filler in paper. The conversion of many paper mills from acid to neutral or alkaline papermaking has led to a much greater use of calcium carbonate, which is now the dominant filler. Both ground and precipitated calcium carbonate are used as filler. The development of onsite calcium carbonate precipitators at paper mills has further eroded the use of kaolin as a filler. However, there still is a fairly large tonnage of kaolin used annually as filler in paper.

Table 15. Properties of a perfect filler

1	Reflectance of 100% at all wavelengths of light
2	High index of refraction
3	Grit-free and a particle size close to 0.3 μm, approximately half the wavelength of light
4	Low specific gravity, soft, and non-abrasive
5	Ability to impart to paper a surface capable of taking any finish, from the lowest matte to the highest gloss
6	Complete retention in the paper web
7	Completely inert and insoluble
8	Reasonable in price

Table 16. Filler grades of kaolin

Type	Brightness
Airfloated kaolin	80–81
Whole clay filler	81–85
Water-washed filler	81–86
Delaminated filler	87–89
Calcined kaolin extender	91–95

Paper is filled to extend fiber for cost reduction and to improve several properties including opacity, brightness, smoothness, and printability. The loading levels of filler range from 2% to 8% in newsprint to as high as 30% in some papers. The two most important properties contributed by kaolin as a paper filler are opacity and brightness. Calcined kaolin gives much more opacity to paper than does hydrous kaolin.

A relatively new use of kaolin is as a fiber extender in the manufacture of gaskets for automobile and truck engines. Gaskets were previously formulated using 80–85% asbestos, but health problems associated with asbestos have led to the use of kaolin. The particle size distribution and platy shape of kaolin are important to the reinforcement and seal of gaskets (Bundy, 1993). Also important is the low abrasiveness of kaolin, which minimizes the die-wear as gaskets are precision die-stamped.

Calcined kaolins are used both as a filler and coating pigment, because of their high brightness and good opacity. Calcined kaolins are used as extenders for titanium dioxide, which is an expensive prime pigment used in both paper filling and coating. In many formulations, up to 60% calcined kaolin can replace titanium dioxide without serious loss of brightness or opacity. The cost of titanium dioxide is of the order of 6 times the cost of calcined kaolin products. Fig. 52 is an scanning electron micrograph (SEM) of the surface of a calcined kaolin particle which exhibits hundreds of small mullite crystallites. The calcined kaolin products have brightness ranging from 91% to 96%. The opacity is increased because the kaolin particles are slightly fused together, which increases the light scatter due to air voids in the slightly fused calcined particles. Light scatter promoted by voids can be shown by the Fresnel reflection coefficient, *R*:

$$R = \frac{N_1 - N_0}{N_1 + N_0}$$

where N_1 is the refractive index of the pigment and N_0 is the refractive index of the media. The greater the difference in the refractive indices of the components of a system, the greater is the Fresnel reflection *R*. Air-filled voids have a much lower refractive index than the calcined kaolin. Calcined kaolin grades are normally very fine in particle size, generally 88–96% less than 2 μm. The calcined kaolin is used as an additive to hydrous kaolin in coating colors to increase brightness and opacity, usually in amounts of 20% or less based on the dry weight. Calcined kaolin is also used as a filler in paper.

2. PAINT

Paint is a significant market for kaolin, although it is considerably less than the market for paper coating and filling. About 600,000 tons annually are used worldwide as extender pigments in paint. The largest use is as a pigment extender in water-based interior latex paints. It is also used in oil-based exterior industrial primers. Calcined and delaminated kaolins are used extensively in interior water-based paints. These paints have moderate to high pigment volume concentrations ranging from 50% to 70%. For semi-gloss and high gloss water-based systems, fine particle size kaolins are used, but at less than 50% pigment volume concentration (Bundy, 1993). The particle size of these fine kaolins used in paint is about 98% less than 2 μm. Kaolin contributes to suspension, viscosity, and leveling of paints. The dominant pigment used in paint is titanium dioxide, so as much calcined kaolin as possible is used to extend the TiO_2 in order to reduce cost.

Delaminated kaolins, because of their high aspect ratio and relatively thin plates, give a smooth surface to paint films and a greater sheen. Scrubbability of a paint is improved with calcined kaolin, as is the toughness of the film. Washability, which is the ease with which a stain can be removed by washing, and enamel holdout (the ability of a substance to prevent the entry of an enamel into its interior structure) are promoted by the use of delaminated kaolins in the paint. In flat paints, calcined kaolin gives better hiding power, film toughness, and scrubbability, but gives poor stain resistance. By proper blending of extenders and pigments, paint formulations can be tailored to specific needs.

3. CERAMICS

Ceramics includes a wide range of products in which kaolins are utilized. These include dinnerware, sanitaryware, tile, electrical porcelain, pottery, and refractories. Kaolins and ball clays, which are kaolinitic clays, are both used as major ingredients in many ceramic products. The term ceramic refers to the manufacture of products from earthen materials by the application of high temperatures (Grim, 1962). Ceramics historically goes back to prehistoric times when early man used earthenware in cooking. He learned that he could form shapes with plastic clays and that heat would fix the shape and make them stable in water.

Through time, with the development of modern science, ceramic art has become an engineering profession. The ceramic properties of clay materials are variable depending on the clay mineral composition and such properties as particle size distribution, presence of organic material, and the non-clay mineral composition. The clay mineral composition is the most important factor determining ceramic properties. Kaolinite is the most important clay mineral used in ceramic applications because of its physical and chemical properties that are imparted to ceramic processing and finished products.

The more important properties that kaolin and ball clay impart to ceramics are plasticity, green strength, dry strength, fired strength and color, refractoriness, ease of casting in sanitaryware, low to zero absorption of water, and controlled shrinkage. Shrinkage is an important property because ceramic articles undergo shrinkage at two different points in the manufacturing sequence. During drying, the article will shrink in varying amounts depending on the composition and the percentage of water present. During firing, the ceramic article will further shrink. Therefore, it is important to know both the drying and firing shrinkage. Linear and volume shrinkage can both be measured, although linear shrinkage is more commonly reported (Jones and Bernard, 1972). In the unfired body, both the water of plasticity and shrinkage generally decrease as the particle size increases. In the fired body, the firing shrinkage and water absorption generally decrease, whereas the modulus of rupture (MOR) and fired whiteness generally increase as the particle size increases (Adkins et al., 2000).

Plasticity is defined as the property of a material which permits it to be deformed under stress without rupturing and to retain the shape produced after the stress is removed (Grim, 1962). The measurement of plasticity has been difficult to determine quantitatively. In general, three ways have been used to measure plasticity. One is to determine the amount of water necessary to develop optimum plasticity or the range of water content in which plasticity of the material is demonstrated. Atterberg (1911) proposed that the lower value, called the plastic limit, and the higher limit, called the liquid limit, is the plasticity index. A second method is to determine the amount of penetration of a needle or some type of plunger into a plastic mass of clay under a given load or rate of loading (Whittemore, 1935). Another way is to determine the stress necessary to deform the clay and the maximum deformation the clay will undergo before rupture. Bloor (1957) presented a critical review of plasticity. A Brabender plastigraph can be used to measure the stress limits mentioned above. Recently, Carty et al. (2000) described a high pressure annulus shear cell or HPASC, as a new plasticity characterization technique.

Green strength is measured as the transverse breaking strength of a test bar suspended on two narrow supports in pounds per square inch or kilograms per square centimeter. Green strength has to be adequate for the piece to be handled without bending or breaking. Ball clays, which are finer in particle size than most kaolins, have a higher green strength (Holderidge, 1956).

Drying shrinkage is the reduction in size, measured either in length or volume, that takes place when the clay piece is dried to drive off the pore water and absorbed water. The drying shrinkage is expressed in percent reduction in size based on the size after drying. In the laboratory, the measurement is made on a test bar after drying for a minimum of 5 h at 105°C. The drying shrinkage is related to the water of plasticity. It increases as the water of plasticity increases and also increases as the particle size decreases. Ball clays have higher dry shrinkage than most kaolins. Table 17 shows that drying shrinkage of kaolinite increases dramatically with a decrease in particle size.

Dry strength is the transverse breaking strength of a test bar that has been dried to remove all the pores and adsorbed water. The dry strength of kaolins and ball clays is greater than their green strength. Dry strength is closely related to particle size which indeed is a major controlling factor. Table 18 shows that the finest fraction of kaolinite has a dry

Table 17. Linear drying shrinkage of kaolinites of varying particle size (Harman and Fraulini, 1940)

Particle size (μm)	Linear drying shrinkage (%)
10–20	1.45
5–10	1.89
2–4	2.19
1.0–0.5	2.35
0.5–0.25	2.69
0.25–0.10	3.70

Table 18. Dry strength of kaolinite in relation to particle size (Anonymous, 1955)

Size fraction	psi
Whole clay	243
Coarser than 1 μm	26
1–0.25 μm	88
Finer than 0.25 μm	750

strength about 30 times higher than the coarse fraction. The fine particle ball clays have a high dry strength.

The fired properties of kaolins and ball clays are most important in determining the ceramic application for a particular kaolin or ball clay product. It should be understood that the non-clay mineral components such as quartz, feldspar, and other mineral additives play an important role in determining the firing characteristics. If organic material is present as it is in ball clays, oxidation to destroy the organic material begins at a temperature of about 300°C and is completed at a temperature of about 500°C. At a temperature between 550 and 600°C (Fig. 60), kaolinite is dehydroxylated and the lattice structure of kaolinite becomes amorphous even though the particle shape is largely retained. This amorphous arrangement of the silica and alumina is retained until a temperature of about 980°C is reached. At that temperature, the amorphous mixture of silica and alumina in metakaolin combines to form a new phase. When this new phase forms, an exothermic reaction takes place. There is some dispute about the phase that is formed at this temperature, but most believe the exothermic reaction is caused by the nucleation of mullite (Johns, 1953). Further heating to a temperature of 1200°C results in larger crystallites of mullite, which Wahl (1958) calls secondary mullite. Kaolinite fuses at 1650–1775°C (Norton, 1968). The fired color of

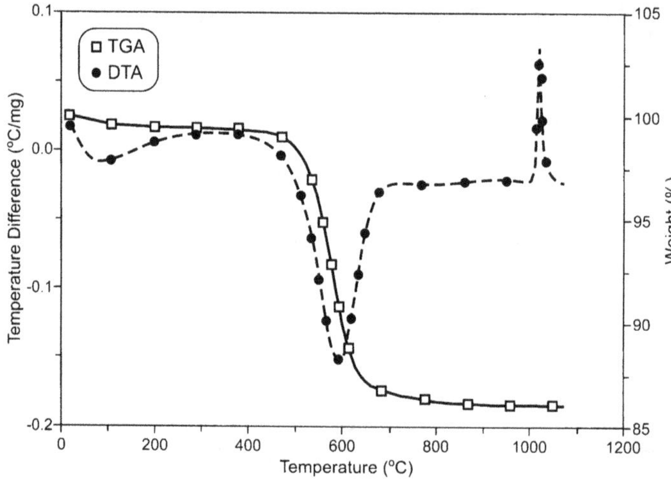

Fig. 60. Typical DTA–TGA curves of kaolinite showing the endothermic and exothermic reactions.

kaolinite is white or near-white. Ball clays fire to a light cream color. The MOR of fired kaolinite and ball clay is very high compared to the MOR of the dried counterparts. The MOR reported for the fired pieces is generally a blend of 50% fine silica and 50% kaolin or ball clay. The MOR ranges from 300 to 900 psi depending largely on the particle size of the kaolin or ball clay.

Casting rate is important in the manufacture of sanitaryware. Fine-grained bodies cast more slowly than coarse ones. The viscosity of a slip must be carefully controlled because if it is too viscous, the slip will not properly fill the mold or drain cleanly and relatively fast. Therefore, viscosity is measured on kaolins and ball clays that are used in the casting process.

Halloysite is used as an additive in the manufacture of high quality dinnerware. The addition of 5–10% by weight in the body provides high fired brightness and increased translucency, both of which are desirable properties of dinnerware.

The use of kaolins and ball clays in refractories began in the early 1800s in New Jersey. Refractory clays are used primarily to make firebricks and blocks of many shapes, insulating bricks, saggers, refractory mortars and mixes, monolithic and castable materials, ramming and air gun mixes, and other refractory products. The specifications for refractory clays are as many as the different uses. Resistance to heat is the most essential property and pyrometric cones are used to indicate the heat duty required. Table 19 shows the values of the pyrometric cones. The pyrometric cone measures the combined effects of temperature and time

Table 19. End points of small Orton pyrometric cones

Cone number	End point (°C)	Cone number	End point (°C)
07	1008	15	1430
06	1023	16	1491
05	1062	17	1512
04	1098	18	1522
03	1131	19	1541
02	1148	20	1564
01	1178	23	1605
1	1179	26	1621
2	1179	27	1640
3	1196	28	1646
4	1209	29	1659
5	1221	30	1665
6	1255	31	1683
7	1264	$31\frac{1}{2}$	1699
8	1300	32	1717
9	1317	$32\frac{1}{2}$	1724
10	1330	33	1743
11	1336	34	1746
12	1355	35	1763
13	1349	36	1804
14	1398		

(Norton, 1968). The cones consist of a series of standardized unfired ceramic compositions molded into the shape of triangular pyramids. The sample of kaolin, ball clay, or the refractory composition is molded into the standard cone shape and is heated along with standard cones so that the end point can be determined in terms of an equivalent cone number (Table 19). Refractory bricks are classed as low, medium, high, and super duty. The pyrometric cone equivalent (PCE) values of low duty are from 15 to 29, medium duty from 29 to $31\frac{1}{2}$, high duty from $31\frac{1}{2}$ to 33, and super duty above 33. Flint clays are very refractory but are non-plastic so are mixed with plastic kaolin and/or ball clays to provide the plasticity needed to form the piece and maintain its shape.

4. RUBBER

Kaolin is used in rubber because of its reinforcing and stiffening properties and it is relatively low cost in comparison with other pigments. In rubber goods which are black, the favored pigment is carbon black, but in non-black rubber goods, kaolin is used (Anonymous, 1955). As mentioned previously, there are hard clays which are fine in particle size and soft clays which are relatively coarse in particle size. Hard clays are used in non-black rubber goods where wear resistance is important. Examples are shoe heels and soles, tires, conveyor belt covers, and bicycle tires. Hard clays give stiffness to uncured rubber compounds which is important in the manufacture of rubber hose, tubing, jar rings, and extruded stocks to prevent sagging or collapsing during manufacture. Hard clay is also used to eliminate mechanical molding troubles in hard rubber goods, household goods, toys, and novelties. Other applications for hard clay in rubber are gloves, adhesives, butyl inner tubes, reclaimed rubber, and neoprene compounds.

When high pigment loadings are used to reduce costs and when abrasion resistance is not particularly important, then soft clays are used. Examples are tire bead insulation, household goods, blown sponges, hard rubber toys, and novelties. Larger amounts of soft clay can be incorporated into the rubber and the extrusion rate is faster than when hard clay is used. Delaminated kaolin with a high aspect ratio is used as a filler in the white sidewall tire because it acts as a barrier to air leakage. Also, there are special surface modified kaolins that are used in order to get better dispersion in the rubber and for improved reinforcement. It is estimated that the total annual tonnage of kaolin used for rubber filler is about 600,000 tons.

5. PLASTICS

Kaolin is used as a filler in plastics because it aids in producing a smooth surface finish, reduces cracking and shrinkage during curing, obscures the fiber pattern when fiberglass is used as reinforcement, improves thermal stability, contributes to a high impact strength, improves resistance to chemical action and weathering, and helps control the flow properties. Filler loading in various plastic compositions varies from about 15% to as high as 60%.

The most important use of kaolin is in polyvinyl chloride (PVC) coatings on wire and cable. Calcined kaolin and silane surface modified kaolin are used to improve electrical resistance and to lower cost. Electrical resistance of PVC is improved by fillers that are hydrophobic. Calcining kaolin to about 1000°C reduces surface energy (Drzal et al., 1983), which develops some hydrophobicity which makes calcined kaolin a preferred filler in PVC. Surface treatment with silanes and other hydrophobic materials further increases the hydrophobicity of the surface.

Generally, the finer the particle size of the kaolin, the better the reinforcement of the physical properties in all polymers. Further improvement in strength can be obtained by virtue of coupling agents, which produce chemical bonding between the kaolin filler and polymer. Fine particle kaolins can substantially increase the impact strength of plastics, for example, in polypropylene and PVC. Bundy (1993) discussed the interaction of silane with the hydroxyl group on the kaolinite surface. The benefit of the silane surface treatment is primarily to improve the dispersion of the kaolin filler. The particle shape of kaolinite as thin platelets benefits some polymers in that flexural modulus, dimensional stability, surface smoothness, and barrier properties (Carr, 1990) are improved.

6. INK

The major inorganic pigment used in ink is kaolin. This is a small but important use of kaolin. Ink formulations are similar to those of paper coating and paint with vehicles (binders) and pigments as the basic components. The most important uses of kaolin in inks are to improve ink holdout and to extend both colored and white pigments (Stoy, 1989). To preserve the gloss of ink films, the kaolin extender must not be so coarse in particle size that the particle protrudes above the film surface. A particle size of the extender kaolin pigment between 0.2 and 0.5 μm is the most effective. Additional requirements for extenders are low abrasion to minimize printing plate wear, easy dispersability, and low oil absorption. Fine particle size water-washed kaolins meet these requirements. For ink to give high print gloss, the ink vehicle must hold out on the paper surface. The platelet shape of kaolinite reduces film permeability, which aids in the holdout of the vehicle on the surface. Fine particle size delaminated kaolins do this most effectively.

Gravure printing represents the biggest ink market for kaolin. Gravure inks can accept relatively large amounts of kaolin, 5–15% on the weight of the resin. Kaolin extends the colorant, sharpens dot formation by imparting thixotropy, and improves holdout. Kaolin is used in low viscosity inks for flexographic printing. Some kaolin is used to extend colorants and to provide thixotropy in inks used for offset printing. Silk screen printing uses relatively thick ink films, that can tolerate the larger particles of calcined kaolin, which provides better light scatter to improve opacity and whiteness. Surface modification of kaolin to make it hydrophobic increases its usefulness in oil-based inks.

7. CATALYSTS

The most important mineral used in the manufacture of carriers for catalysts is kaolin. The largest use of kaolin is in catalyst substrates in the catalytic cracking of petroleum. Because many catalysts are used at high temperatures and pressures, the refractory character of kaolin is appropriate for these applications. The purity of the kaolin is critical in this petroleum cracking operation so a processed kaolin with low iron, titanium and alkali, and alkaline earth compounds is preferred.

Kaolin is converted to a zeolite in the preparation of the cracking catalyst supports. This conversion of kaolin to zeolite increases the surface area of the catalyst that is exposed in the reaction. Hettinger (1991) stated that the low cost, high purity, and platelet shape promotes the formation of good pore structure, ease of acid leaching, and ease of conversion to zeolite make kaolin the preferred clay used for catalyst carriers. It is estimated that over 200,000 tons of kaolin are used annually to produce petroleum cracking catalysts (Hettinger, 1991).

Automotive exhaust emissions are controlled by catalytic converters located in the exhaust system. Oxidation catalysts in the converter convert the carbon monoxide and other gases and hydrocarbons produced from incomplete combustion into carbon dioxide and water. Catalytic materials in the converter are supported on a ceramic honeycomb monolith (Fig. 61). This honeycomb contains 46–62 square channels per square centimeter and each channel is coated with an activated alumina layer called a washcoat. Platinum, palladium, and rhodium metal catalysts are dispersed in the washcoat. The ceramic monolith is cordierite, which has a very low coefficient of expansion, so can withstand continued heating and cooling cycles (Murray, 1994). The raw materials used to make the cordierite monolith are calcined kaolin, calcined talc, alumina, and hydrous kaolin. Cordierite ($Mg_2Al_4Si_5O_{18}$) is comprised of 13.7% MgO, 34.9% Al_2O_3, and 51.4% SiO_2. Fig. 62 shows the tertiary diagram. The kaolin must be very plastic and have a high green and dry strength (Murray, 1989).

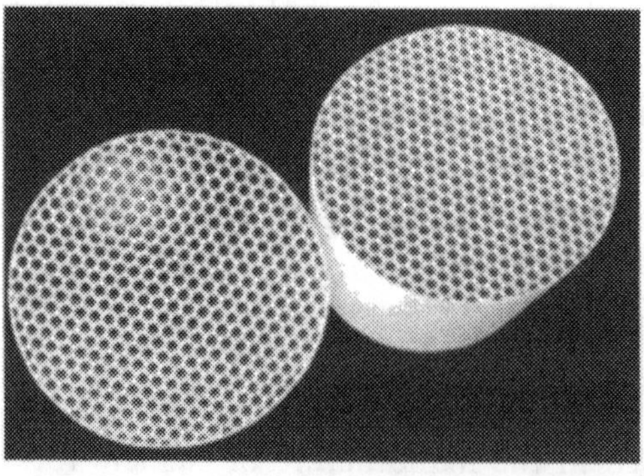

Fig. 61. Catalytic converter honeycomb.

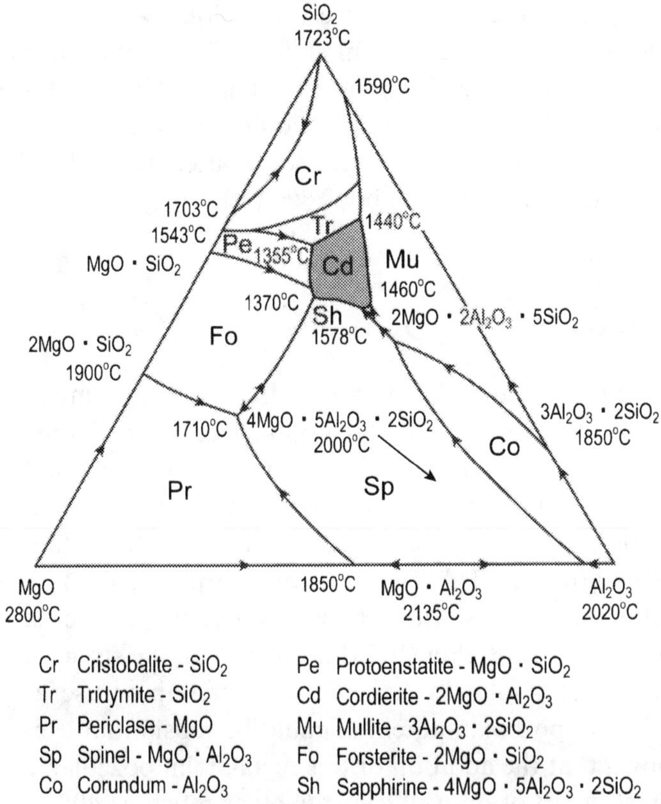

Fig. 62. Temperature and composition to form cordierite.

Kaolin and halloysite are used to make cracking catalysts, as a polymerization catalyst, peptide bond formation, and others (Van Olphen, 1977). Halloysite and metakaolin are used in the manufacture of molecular sieves used as petroleum cracking catalysts. The kaolin for this application must be low in iron and in alkalies and alkaline earth elements. The dry surface of kaolinite is very acidic (Solomon and Murray, 1972) and is used to promote the polymerization of styrene, heterolytic breakdown of organic peroxides, dehydration of alcohols, hydrolysis of esters, and isomerization of alkenes (Solomon et al., 1971).

8. FIBERGLASS

Kaolin is a major component used in the production of fiberglass. Fiberglass has a large number of applications, including insulation, reinforcement of plastics, textile yarn, electronic circuit board substrates, paper, cloth, and roofing shingles. The basic component materials used to make fiberglass are silica, kaolin, and limestone, along with small amounts of boric acid, soda ash, and sodium sulfate. The kaolin must meet rather stringent chemical specifications (Watkins, 1986): Al_2O_3 38.5±0.6%; SiO_2 45.0±0.5%; TiO_2 1.5±0.3%; Fe_2O_3 0.6% maximum. A sizeable tonnage of kaolin, which is dry processed, is used annually in this market and the estimated tonnage is about 800,000 tons.

9. PORTLAND CEMENT

Cement is made by mixing materials containing lime, silica, alumina, and iron oxide. This mixture is sintered and then pulverized at which time a retardant, gypsum is added. Kaolin is an ideal source of alumina and silica and also makes the cement whiter. Relatively recently, a metakaolin (partially calcined) product is used as a pozzolanic additive in certain cements where high strength is needed. The reactive amorphous alumina and silica in the metakaolin reacts with excess calcium to produce a calcium aluminum silicate which is elongate, thus increasing the strength of the concrete. The use of kaolin in cement is a very minor use at present, but the metakaolin pozzolan potential could be substantial. Recently, it has been shown that the addition of this metakaolin pozzolanic material increases the strength of oil well cements by as much as 40%.

10. MISCELLANEOUS USES

There are a multitude of uses of kaolin which are briefly described. The kaolin surface is acidic, but as shown by Solomon and Murray (1972), the acidic surface of kaolinite is largely neutralized in the presence of water, but dry surfaces show a large increase in acidity. At 1% surface moisture, the acidity is equivalent to 48% sulfuric acid and at approximately 0% moisture surface acidity is equivalent to 90% sulfuric acid. This surface acidity and activity must be considered in many of its uses which enhance some and is deleterious in others.

The properties which make kaolin useful in many of the miscellaneous uses are its fine particle size, white color, platy shape, chemical composition, absorbency, low abrasiveness, surface activity, hydrophilic surface which can be easily converted to be organophyllic or hydrophobic, low dielectric constant, low heat conductivity, ease of dispersion, and low viscosity at high solids concentration.

10.1. Alum

In the process to make alum the hydrous kaolin is heated to about 650–700°C, which forms metakaolin, an amorphous mixture of alumina and silica. The metakaolin is reacted with sulfuric acid to produce alum ($Al_2(SO_4)_3 \cdot H_2O$). To a much lesser extent, kaolin is reacted with phosphoric acid to produce aluminum phosphate.

10.2. Abrasive Wheel Bonding

Plastic and refractory kaolins are used to bond the abrasives used in an abrasive wheel. The kaolin and the abrasive are mixed, formed, dried, and fired. The drying and firing shrinkage must be low in order to prevent shrinkage cracks in the abrasive wheel or an abrasive bar.

10.3. Adhesives, Sealants, and Caulks

Adhesive, sealant, and caulk products are used in a large number of end uses and several industrial minerals are employed as fillers, extenders, and pigments. The mineral content in adhesive formulations ranges from 10% to 70%. Kaolin is incorporated into some formulations to improve adhesion, lower drying time, and to increase viscosity. Calcined kaolin, because the hydroxyls on the surface have been eliminated, gives low moisture pickup which provides excellent performance in moisture sensitive sealant applications. The platy shape and white color are important properties in some adhesives and sealants.

10.4. Cosmetics

Cosmetics serve a luxury market and the products have a high added value. Baby powder and body powder attractively packaged may sell for over a hundred times the cost of the kaolin or talc, which make up the powder with a scent additive. Because of the association of talc with fibrous asbestiform minerals, delaminated kaolins have replaced a large portion of the talc which was used formerly. Sterilized delaminated kaolin can make up 75% of a body powder formulation. The particle size of the kaolin is very fine and grit-free. It has good covering power, excellent grease resisting properties, and has good adhesion to the skin. The other major use of kaolin in cosmetics is in face packs and masks. Up to 5% of the formulation can be a fine, particle size kaolin. A recent use of kaolin is in the formulation of a hair conditioner. The kaolin adds body to fine hair, which increases the apparent hair volume.

10.5. Crayons and Chalk

Fine particle size, grit-free kaolin is often used to stiffen crayons and make them more resistant to bending at higher temperatures. Also, the kaolin helps disperse the dye or other organic colorants that are used. The platy fine particle kaolin promotes a smooth thin surface coating on paper or other medium which are colored. Kaolin is also used in chalk because of its softness, plasticity, binding power, and volume.

10.6. Enamels

Porcelain enamels are glassy coatings fused onto metals to provide corrosion protection and decoration. Enamel producers formulate special glasses called frits, which are the major constituent in enamels. The frits are compounded to meet color, opacity, chemical resistance, and processing requirements of the user. The enamel is applied in a thin coat and fused permanently to the metal surface in a low temperature furnace. The frit is compounded to include a number of finely pulverized ingredients which are oxides of coloring agents, whiting, feldspar, kaolin, ball clay, borax, and finely ground glass with a low melting temperature. Kaolin and ball clay are used because of their suspending power at high solids in water and to enhance the dispersion of all the ingredients. The kaolin and/or ball clay used for this purpose must be fine grained and have a white or near-white color when the enamel is fused to the metal surface at low temperature.

10.7. Fertilizers

Kaolins are used as additives to chemical fertilizers as diluents to provide the optimum relative concentration of elements. Kaolins are also used as prilling materials to coat particles of ammonium nitrate, a major component in many fertilizers. The ammonium nitrate particles are deliquescent and become sticky. A thin coating of kaolin makes the ammonium nitrate prills free flowing.

10.8. Fluoride Absorption

Kaolinite has an affinity for fluoride which reacts and perhaps replaces hydroxyls in the structure. If drinking water contains high fluoride levels, kaolinite is used to remove or lower the fluoride content by absorption. A potential new application is in scrubbers to reduce fluorine emissions in some ceramic and clay plants.

10.9. Food Additives

Kaolins are non-toxic and contain little or no deleterious metal ions. It is approved for internal use by the pure food and drug administration. Therefore, it is used in a limited number of foods as an additive. Examples are to stiffen frosting on cakes, added to coatings on chocolate to prevent melting, as a dusting agent in sugar to improve the adherence of the sugar on doughnuts (Rosner, 1958), and for emulsifying certain liquid foods.

10.10. Foundry

Plastic clays which are kaolinitic are widely used in bonding molding sands when a relatively high refractoriness is required, particularly when a molten metal is poured which has a high temperature. Ball clays with a high plasticity are commonly used. These fine particle kaolinitic clays have a lower bond strength than montmorillonite clays. In some very high temperature, molten metal foundries require that sand size granules of kaolin calcined at temperatures of about 1300°C to form mullite are used instead of silica sand.

10.11. Fruit and Vegetable Protection

A relatively new application of kaolin is in spray-coating apples, olives and tomatoes, and other fruits and vegetables to protect them from sun damage as they ripen. A thin coating protects the fruit and vegetable from sun damage by absorbing ultraviolet rays. Rain will wash off the coating, which requires that the fruit or vegetable be spray coated again. Insecticides can be added to the coating to protect the fruit or vegetable from insect damage (Martin, 2002).

10.12. Insecticide and Pesticide Carriers

Some kaolin products with a very fine particle size are used as carriers of insecticides and pesticides. In most applications, the moisture content must be less than 1%. The clay surface must be compatible chemically with the active ingredients to avoid deterioration or breakdown of the chemical with the resulting loss of potency. In some applications, the kaolin is used as a diluent. As described above, kaolin is treated with selected pesticides and/or insecticides and is sprayed as a slurry onto fruit trees and other garden products. Many pesticides are in concentrated form, which can have a harmful effect on plants and must be diluted for effective and economical application.

10.13. Medicines and Pharmaceuticals

Kaolins are used as an absorptive for gastro-intestinal disorders, as a tablet or capsule diluent, as a suspending agent, in poultices and for dusting in surgical operations (Russel, 1988). As an absorptive, clays absorb toxins and harmful bacteria in addition to forming a soothing protective coating on inflamed mucous membrane in the digestive tract (Goodman and Gilman, 1955). Kaolins used in medicines and pharmaceuticals must be free of toxic metals, grit, and be sterilized to remove pathogenic micro-organisms. Kaolin is used as a suspending agent for pectins in the well-known product kaopectate. Kaolin is also commonly used as a diluent in capsules and tablets. In tablets, it aids in making the tablet strong and dense when the tablet is compressed.

10.14. Pencil Leads

Fine particle kaolin is used along with a minor amount of bentonite to bond graphite in pencil leads (Murray, 1961). The graphite and plastic kaolin are mixed and extruded to form the pencil lead. The lead is dried and fired to produce a strong pencil lead. The hardness of the lead, 2 H, 3 H, 5 H, etc. is controlled by the percentage of clay in the lead. A soft lead 2 H contains less clay than a harder 5 H lead.

10.15. Plaster

Kaolins are used in plaster as a white colorant, to disperse and improve the uniformity of the plaster, to increase the percent solids and reduce the water content, and to improve the workability and flowability. Fine particle size kaolin is preferred for this use.

10.16. Polishing Compounds

Ultra-fine calcined kaolin is used in many polishing compounds. The particle size is 100% finer than 3 μm and 90% finer than 2 μm. Calcined kaolin has a hardness of between 6 and 7 on the Mohs' hardness scale. This product is used in toothpaste, automobile polishes, polishes for silver and gold, which are soft metals and require a mild polishing action which removes the oxidized surface. The calcined kaolin must be free of coarse, abrasive particles, which would cause scratching or gouging. Most automobile polishes contain this fine particle size calcined kaolin as the major polishing agent in the polish.

10.17. Roofing Granules

Granular calcined kaolin is spread on the surface of the asphalt paper used to cover roofs. The calcined kaolin is white so is a good reflector. It is hard, durable, and insoluble, which are properties needed for granules spread on a roof. The granules can be sized to make coarse, medium, or fine products.

10.18. Sizing

Kaolins, generally mixed with an adhesive, are used to coat nylon and other synthetic fibers and also for some cotton goods. Very fine particle size kaolins, less than 2 μm, provide a white color and make the filaments in a spinning yarn more homogenous and better able to withstand the strain and friction of weaving. Another related use of kaolin is in carpet backing. A relatively coarse kaolin is used for this purpose. The major reason for use in carpet backing is to reduce cost as the kaolin is much less costly than the rubberized backing.

10.19. Soaps and Detergents

Kaolins are used in soaps as a partial replacement for the fatty acid component because of their emulsifying action, their affinity for carbon particles, and their detergent affect. In all probability, the kaolin is inert and serves only to dilute the soap and to aid in the dispersion of the fatty acid component. In recent years, much of the phosphate used in detergents has been replaced by synthetic zeolites. Zeolites can easily be prepared from kaolin by reacting the kaolin with sodium, calcium, or magnesium hydroxide at a temperature of about 100°C. A pressure vessel will speed up the reaction. A low iron kaolin is preferred for this use.

10.20. Tanning Leather

Kaolins are used in the tanning of leather to lighten the color and to give the leather a softer and smoother feel. A fine particle size kaolin is necessary as the fine particles can readily penetrate the leather and fill the pores.

10.21. Welding Rod Coating

Kaolin, especially metakaolin, has a high dielectric constant and is used to coat welding rods. This coating keeps the electric current moving to the top of the welding rod so it will melt and provide a molten metal fusion.

10.22. Wire Coating

Metakaolin is used to fill the plastic- or rubber-coating material on wires that carry an electric current. The high dielectric constant of the metakaolin in the coating contains the electric field in the wire. This is a sizeable market for metakaolin.

REFERENCES

Adkins, T., et al. (2000) Kaolin particle size distribution effects on whitewares—related performance properties. Chapter in *Science of Whitewares*. Carty, W.M. and Sinton, C.W. eds. American Ceramic Society, Westerville, OH, pp. 121–130.

Anonymous (1955) *Kaolin Clays and their Industrial Uses*. J.M. Huber Corp., NY, 214pp.

Atterberg, A. (1911) Die plastizitat der tone. *Int. Mitt. Bodenk,*, I, 4–37.

Bloor, E.C. (1957) Plasticity: a critical survey. *Trans. Brit. Ceram. Soc.*, 56, 324–481.

Bundy, W.M. (1967) Kaolin properties and paper coating characteristics. *Chem. Farg. Prog.*, 63, 57–67.

Bundy, W.M. (1993) *The Diverse Industrial Applications of Kaolin*. Special Publication No. 1, Clay Minerals Society, Boulder, CO, pp. 43–73.

Bundy, W.M. and Ishley, J.H. (1991) Kaolin in paper filling and coating. *Appl. Clay Sci.*, 5, 397–420.

Carr, J.B. (1990) Kaolin reinforcements: an added dimension. *Plast. Compound.*, September/October, 108–118.

Carty, W.M., et al. (2000) Plasticity revisited. Chapter in *Science of Whitewares*. Carty, W.M. and Sinton, C.W. eds. American Ceramic Society, Westerville, OH, pp. 225–236.

Drzal, Z., et al. (1983) Effects of calcination on the surface properties of kaolinite. *J. Colloid Interf. Sci.*, 93, 126–139.

Goodman, L.S. and Gilman, A. (1955) *The Pharmacological Basis of Therapeutics*, 2nd Edition. MacMillan Co., NY.

Grim, R.E. (1962) *Applied Clay Mineralogy*. McGraw-Hill, NY, 422pp.

Harman, C.G. and Fraulini, F. (1940) Properties of kaolinite as a function of its particle size. *J. Am. Ceram. Soc.*, 23, 252–298.

Hettinger, W.P. Jr. (1991) Contribution to catalytic cracking in the petroleum industry. *Appl. Clay Sci.*, 5, 445–468.

Holderidge, D.A. (1956) Ball clays and their properties. *Trans. Brit. Ceram. Soc.*, 55, 369–440.

Johns, W.D. (1953) High temperature phase changes in kaolinite. *Miner. Mag.*, 30, 186–198.

Jones, J.T. and Bernard, M.E. (1972) *Ceramics: Industrial Processing and Testing*. Iowa State University Press, Ames, IA, 213pp.

Lagaly, G. (1989) Principles of flow of kaolin and bentonite dispersions. *Appl. Clay Sci.*, 4, 105–123.

Malla, P.B. and Devisetti, S. (2005) Novel kaolin pigment for high solids ink jet coating. *Paper Tech.*, 46(8), 17–27.

Martin, C.C. (2002) Personal communication.

Murray, H.H. (1961) *Pencil Clays*. US Patent 2986472.

Murray, H.H. (1975) Applied rheology. *Proc. Porcelain Enamel Inst.*, 37, 1–9.

Murray, H.H. (1989) Clay minerals for advanced ceramics. *Mining Eng.*, 41, 1123–1126.

Murray, H.H. (1994). Catalysts. Chapter in *Industrial Minerals and Rocks*, 6th Edition. Carr, D.D., ed. Society for Mining, Metallurgy and Exploration, Littleton, CO, pp. 191–193.

Murray, H.H. and Kogel, J.E. (2005) Engineered clay products for the paper industry. *Appl. Clay Sci.*, 29, 199–206.

Norton, F.H. (1968) *Refractories*, 4th Edition. McGraw-Hill, NY, 228pp.

Pickering, S.M. Jr. and Murray, H.H. (1994) Kaolin. Chapter in *Industrial Minerals and Rocks*, 6th Edition. Carr, D.D., ed. Society for Mining, Metallurgy and Exploration, Littleton, CO, pp. 255–277.

Rosner, C.J. (1958) Manufacture of Improved Doughnut Sugar and the Resulting Product. US Patent 2,846,311.

Russel, O. (1988) Minerals in pharmaceuticals, the key is quality assurance. *Ind. Miner.*, August, 32–43.

Solomon, D.H. and Murray, H.H. (1972) Acid–base interactions and the properties of kaolinite in non-aqueous media. *Clay. Clay Miner.*, 20, 135–141.

Solomon, D.H., et al. (1971) The quality of clay minerals in polymerizations and related reactions. *J. Macromol. Sci. Chem.*, 3, 587–601.

Stoy, W.S. (1989) Make room for extenders. *Am. Ink Maker*, June, 46–50.

Van Olphen, H. (1977) *An Introduction to Clay Colloid Chemistry*, 2nd Edition. John Wiley and Sons, NY.

Wahl, F.M. (1958) *Reactions in Kaolin-Type Minerals at Elevated Temperatures as Investigated by Continuous X-Ray Diffraction*. PhD Thesis, University of Illinois.

Watkins, E.C. (1986) Mineral raw materials for fiberglass manufacturing. Society for Mining, Metallurgy and Exploration preprint, *New Orleans Annual Meeting*, 5pp.

Whittemore, J.W. (1935) Mechanical method for measurement of plasticity of clay. *J. Am. Ceram. Soc.*, 18, 352–360.

Willets, W.R. (1958) *Paper Loading Materials*. Monograph 19, Tappi, New York, p. 5.

Wilson, I.R. (2004) Special clays. *Ind. Mineral. Mag*. November, 54–61.

Yuan, J. and Murray, H.H. (1997) The importance of crystal morphology on the viscosity of concentrated suspensions of kaolins. *Appl. Clay Sci.*, 12, 209–219.

Chapter 6

BENTONITE APPLICATIONS

As discussed previously, bentonite is a rock term. Bentonites are comprised predominantly of the smectite group of minerals. Table 20 shows the clay minerals that make up the smectite group. The most common are sodium and calcium montmorillonites. Calcium montmorillonite is the most predominate of the smectite minerals and is found in many areas of the world. Sodium montmorillonite is relatively rare in occurrence in comparison with calcium montmorillonite. The largest and best-known occurrence is in the states of Wyoming and Montana in the United States. Saponite occurs in a few areas of the world and hectorite, beidellite, and nontronite are rare. Nontronite occurs mainly in iron-rich soils. Volkonskoite and sauconite are extremely rare and may occur in only one or two locations. Beidellite is the aluminum montmorillonite and is also relatively rare in occurrence.

The smectite minerals occur as extremely fine particles of the order of 0.5 μm or less (Fig. 11). Exchangeable cations such as sodium, calcium, and magnesium occur between the silicate layers, associated with water molecules. These elements are exchangeable and the property of exchange capacity is measured in terms of milliequivalents per 100 grams. The property of ion exchange and the exchange reaction are very important in many of the applications in which the smectite minerals are used. For example, in soils, plant foods are frequently held in the soils as exchangeable ions. The cation exchange capacity of smectites range from about 40 in calcium montmorillonite to 150 milliequivalents in hectorite

Table 20. Smectite clay minerals

Sodium montmorillonite
Calcium montmorillonite
Saponite (Mg)
Beidellite (Al)
Nontronite (Fe)
Hectorite (Li)
Volkonskoite (Cr)
Sauconite (Zn)

per 100 grams. Sodium montmorillonite has an exchange capacity which generally is between 80 and 110.

The water molecules that occur between the layers in smectites are called low temperature water which can be driven off by heating from 100 to 150°C (Grim, 1968). It has been shown that the water on the surface between the montmorillonite layers is in a physical state different from liquid water (Low, 1961). A multitude of studies of this water between the layers indicate that the water molecules are structurally oriented to form an ice-like structure (Bradley, 1959). Johnson et al. (2005) used infrared absorption to provide new information about the clay water interface and the role of exchangeable cations. The thickness of these water molecules between the montmorillonite layers is related to the exchangeable cation present. When sodium is the exchangeable ion, the water layer is about 2.5 Å, which is one water layer and when calcium or magnesium is the exchangeable cation, then the layer is about 4.2–4.5 Å thick, which is two water layers. A sodium montmorillonite has a layer spacing of about 12.5 Å and a calcium montmorillonite layer has a spacing of 14.2–14.5 Å.

In the octahedral layer of the smectites in which all three octahedral positions are filled is called trioctahedral and when only two-thirds of the possible positions are filled is called dioctahedral. An example of a trioctahedral smectite is saponite when Mg^{++} fills all the octahedral positions. Beidellite is an example of a dioctahedral smectite when Al^{+++} fills only two out of three octahedral positions.

The color of smectites can vary from tan to brown to brownish green or blue green and is rarely white. Color controls the use in some cases. Some important properties of smectites that relate to their applications are shown in Table 21. For the sodium montmorillonites important properties related to their use are viscosity, swelling capacity, thixotropy,

Table 21. Important physical and chemical properties of smectites

2:1 Expandable layers
High layer charge
High base exchange capacity
Very thin flakes
High surface area
High absorption capacity
High swelling capacity
High viscosity
Thixotropic
Color: tan, olive green, brown, blue-gray, white

Table 22. Applications of smectites

Drilling muds	Dessicants	Pharmaceuticals
Foundry bonds	Detergents	Pillared clays
Iron ore pelletizing	Emulsion stabilizers	Plasticizers
Cat litter	Fertilizer carrier	Rubber filler
Absorbents	Food additive	Sealants
Adhesives	Fulling wool	Seed growth
Aerosols	Herbicide carrier	Soil stabilization
Animal feed bonds	Industrial oil absorbent	Slurry trench stabilization
Barrier clays	Insecticide and pesticide carrier	Suspension aids
Bleaching earths	Medicines	Tape joint compounds
Catalysts	Nanoclays	Water clarification
Cement	Organoclays	
Ceramics and refractories	Paint	
Cosmetics	Paper	
Crayons	Pencil leads	
De-inking newsprint		
Deodorizers		

impervious filter cake, and dispersability. For the calcium montmorillonites important properties related to their use are high absorption capacity, bonding strength, and bleaching capability. Table 22 shows the multitude of uses of the smectites (Kendall, 1996).

As mentioned in Chapter 5, the physical and chemical properties of smectites are very different from kaolinite. The most significant differences compared with kaolinite relate to their structure and composition and their very fine particle size, relatively high base exchange capacity, high surface area, high viscosity and swelling capacity, and high absorptive capacity. It is these different physical and chemical properties that account for many of the significantly different applications of smectites compared with kaolins. Also, sodium and calcium montmorillonites have significantly different properties which accounts for some of their unique uses. Sodium bentonites are noted as high swelling clays and calcium bentonites as low swelling clays.

1. DRILLING FLUIDS

Sodium montmorillonite (Na bentonite) is the major constituent of freshwater drilling muds. The function of the drilling mud is to remove cuttings from the drill hole to keep formation fluids from penetrating into the drilling mud, to lubricate and cool the bit, and to build an impervious filter cake on the wall of the drill hole to prevent the penetration of water from the drilling fluid into the formations and formation fluids from the drilling mud. High viscosity is required in order to remove the cuttings from the hole. The circulating drilling fluid carries the cuttings up the hole and removes them by screening (Fig. 63). Another important quality besides high viscosity is that the mud must be thixotropic. This thixotropic property is when the drilling ceases, the mud must rapidly form a gel to prevent the cuttings from settling to the bottom of the drill holes and freezing the bit so that the drill stem breaks. The second important thixotropic property is when the drill starts again, the drilling mud must become fluid. Sodium bentonite has this thixotropic property and the western bentonite is widely used in drilling fluids all over the world. Also,

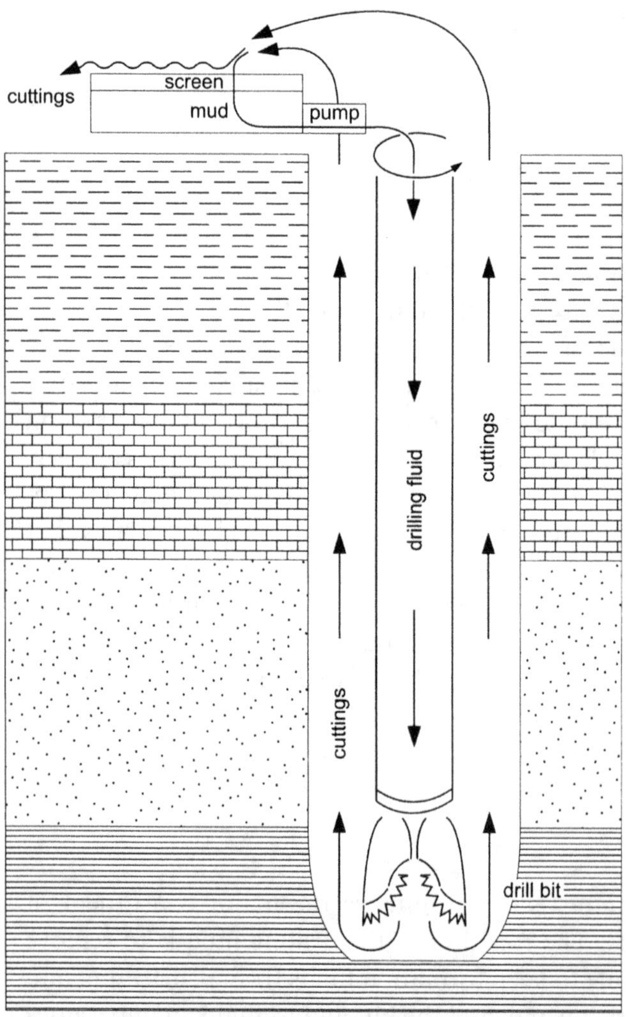

Fig. 63. Schematic showing drilling mud flow in an oil well.

it forms a thick impervious cake along the edge of the hole, which prevents the drilling fluid from penetrating porous formations. The American Petroleum Institute sets the specifications for bentonite that is used in drilling oil wells. Many of the Wyoming and Montana sodium bentonites meet the American Petroleum Institute (API) specifications. The sodium bentonite gives mud yields of over 100 bbl/ton. A 5% addition of the sodium bentonite usually gives the desired viscosity. This bentonite has a high gel strength and a low filter cake permeability all of which make these western bentonites the premier drilling mud in the world.

2. FOUNDRY BONDS

Molding sands composed of silica sand and bentonite are used extensively in shaping metal in the casting process. Bentonite is used to provide the bonding strength and plasticity to the sand–clay mixture. Tempering water is added to the mixture to make it plastic and cohesive so that it can be molded around a pattern. The tempering water is a small percentage of the mix, usually about 5%. The sand–clay mix must be strong enough to maintain the molded shape after the pattern is removed and while the molten metal is being poured into the mold.

The important properties of the sand–clay mix are green compression strength, dry compression strength, hot strength, flowability, and permeability. Green compression strength is the compressive force necessary to cause failure in a test specimen containing tempering water and bentonite compacted by ramming. Dry compression strength is the compressive force necessary to cause failure in a rammed specimen that has been dried to remove all the tempering water. Hot strength is the compressive force necessary to cause failure of a rammed test specimen at a high temperature. The high temperature is generally of the order of 1100°C. Flowability is the property that permits the sand–clay mixture to fill recesses that may be present in the pattern. Good flowability may require that the amount of tempering water be considerably higher than that required for maximum green strength (Grim and Johns, 1957). Permeability is measured on the green or dry test specimens. This property is important because it allows any gas present in the molten metal to escape through the mold. Other properties that are important are bulk density, durability, ease of shake out of the sand–clay mold from the casting, and cleanness of the surface of the cast metal after shake out. These latter three properties can only be determined after the sand–clay mixture is actually used in foundry practice.

Both sodium and calcium montmorillonites are used as bonds for the foundry sand. Each of these montmorillonites have different properties and in many cases, blends of these two bentonites provide the optimum properties that are needed in a particular foundry. Calcium bentonite has a higher green strength, lower dry strength, lower hot strength, and better flowability than sodium bentonite. Methods of testing sand–clay mixtures are outlined in publications of the American Foundrymen's Society (Anonymous, 1963).

3. PELLETIZING IRON ORE

Sodium bentonite is used to pelletize iron ore (Devaney, 1956). Finely pulverized iron ore concentrates are pelletized into marble-sized spheres about 2.5 cm in diameter for ease of handling and shipping and to produce a superior furnace feed. The sodium bentonite constitutes about 0.5 wt.% of the pelletized ore. The reason that sodium bentonite is the preferred clay for pelletizing is its superior dry strength and the low percentage necessary to bind the pellet. The mixtures of bentonite, iron ore, and water are commonly tested for wet drop strength, wet compression strength, plastic deformation, and dry compression strength.

4. CAT LITTER

Both calcium and sodium bentonites are used as constituents in cat litter, but for very different reasons. Calcium bentonite is used because of its high absorbent quality. The calcium bentonite is dried, crushed, and sized into a granular product for use as a litter box filler. In recent years, a new litter called clumping cat litter has become the preferred type of litter. This litter is made by blending high swelling sodium bentonite with the calcium bentonite granules. When the feline waste hits this blend of granules, the sodium bentonite swells and forms a hard clump which is easy to remove from the litter box. This saves dumping the litter and also keeps down the odor. Currently, this is the highest annual tonnage use of sodium bentonite.

5. ABSORBENTS

Calcium bentonites are very good absorbent clays. This is because of their surface charge and surface area. Many of the calcium bentonites will absorb up to 100% of their dry weight of water and up to about 80% of their weight of oil. Most calcium bentonites marketed for use as an absorbent are produced in granular form. The American Society for Testing and Materials (ASTM) standard is C431-65, which is titled *Standard Methods for Sampling and Evaluation of Sorptive Mineral Products*. Attrition resistance is an important property and is measured by shaking the granules with steel balls on a screen according to a specified procedure as mentioned above.

6. MISCELLANEOUS APPLICATIONS

6.1. Adhesives

Bentonites are used in a variety of adhesives including lignins, starch, latex, and asphalt. The major uses are in adhesives for paper products and in cements for floor coverings such as linoleum, rubber, and asphalt tile. Bentonites used in adhesives are not always inert diluents but can provide improved properties. These include reduction in the penetration of the adhesive into the articles to be joined, increased solids content of the adhesive, faster setting rates, and superior bond strength. The dispersion and suspension characteristics of montmorillonite make it particularly useful in adhesives made with latex and asphaltic materials and also in starch, casein, and sodium silicate adhesives.

6.2. Aerosols

In some aerosols, very fine particle size sodium montmorillonite is used as a carrier for the ingredient such as a mosquito repellent. It also keeps the repellent on the arms or legs or wherever the repellent is sprayed for a longer period of time. White bentonite is preferred for this application.

6.3. Animal Feed Binders

Both sodium and calcium bentonites are used to bind animal feed into pellets (Saeed, 1996). The finely pulverized bentonite is very plastic and binds the feed and other necessary medicinal feed supplements such as antibiotics, vitamins, and minerals, into pellets which are easy to package and handle. The sodium and calcium montmorillonites act as absorbents for bacteria and certain enzymes, which when removed from the animal, promotes faster growth and better health. In the production of the feed pellets, the bentonite reduces friction and adhesion in the pellet extruder.

Data from several studies show that binding the feed pellets with bentonite improves the feed efficiency by increasing the weight gain in swine and cattle, increased egg production from chickens, and increased milk yield from dairy cattle (Kendall, 1996). Also, recent studies at Texas A&M have shown that the bentonite binds to aflatoxin and mycotoxin preventing their uptake in the animal's stomach (Kannewischer et al., 2005).

6.4. *Barrier Clays*

Sodium bentonites are used extensively for water impedance because of their high swelling capacity. The high swelling sodium bentonite swells and fills the pores and voids in the material into which it is incorporated preventing water or other liquids from moving through the barrier. Common uses are in earthen structures such as dams, to seal irrigation ditches, to prevent seepage of water from ponds and impounds, and to prevent water from entering basements of homes. Sodium bentonite is also used in landfills and toxic waste dumps as liners to prevent water from entering and liquids from exiting (Keith and Murray, 1994).

6.5. *Bleaching Earths*

Calcium bentonite, known as fuller's earth in Great Britain (Robertson, 1986), is acid activated to make bleaching earths used as a refining and clarifying agent in the production of edible oils and fats, and industrial oils and waxes. Acid activation enhances properties already present in the clay by changing certain chemical and physical attributes without destroying the crystal structure. Sulfuric acid is most commonly used in the activation process but hydrochloric acid is also effective. The acid dissolves some impurities such as calcite and gypsum, replaces exchangeable divalent calcium and magnesium ions with monovalent hydrogen ions, and dissolves some aluminum ions from the tetrahedral layer and some iron, aluminum, and magnesium ions from the octahedral layer. This acid treatment increases the charge on the lattice and increases the surface area. The acidity of the activated bentonite surface and the increased surface area contributes to the bleaching activity in the refining process. The refining removes a variety of impurities including phosphatides, fatty acids, gums, trace metals, and absorbs organic color bodies which bleaches the oil and also deodorizes the oil. Acid-activated calcium bentonites are used to refine and bleach palm oil, animal fats, coconut, soybean, rapeseed, sunflower, and corn oils (Griffith, 1990). It is estimated that over 800,000 tons of bleaching clays are used annually and in the Asian countries of China and India, the use is rapidly expanding.

6.6. Catalysts

Calcium montmorillonites are used in some processes involving the catalytic cracking of petroleum (Hettinger, 1991). Sodium montmorillonite is used for the dehydration of oils such as castor oil. Montmorillonites or acid-treated montmorillonites have been used for numerous reactions including the dimerization of unsaturated fatty acids to dicarboxylic acids, the alkylation of phenols, and for many laboratory syntheses, for example, the preparation of di-2,2′-alkyl ethers. Montmorillonite is also used in the manufacture of polystyrene and similar compounds and in the synthesis of terpenes. Cation exchanged montmorillonites are also effective catalysts including Ni montmorillonite for purification and hydrogenation of edible fats, Al and Cr montmorillonites for lactonization reactions, and Fe and Co montmorillonites to protonate several organic species. Ion exchanged montmorillonites behave as solid acid catalysts and are effective and selective catalysts for the hydration of ethylene, aluminum exchanged montmorillonite was the most effective. Sodium bentonite exchanged with cations of high charge density such as Al, Cu, Fe, and Cr is an efficient and selective catalyst for the production of ethyl acetate from ethylene and acetic acid. There are many patents in the catalysis field using montmorillonites as the template.

6.7. Cement

The addition of 1–2% sodium bentonite to Portland cement in concrete and cement slurries improves workability, lessens aggregate segregation, and improves the impermeability. Mielenz and King (1955) reported that bentonitic shales were used in the preparation of pozzolans. Recent studies have shown that bentonites are an acceptable pozzolan in cement.

6.8. Ceramics

Bentonites are not a major component in ceramic products, but are in many cases, an important additive. In some brick clays, there is a lack of plasticity and a small addition of sodium or calcium bentonite will improve the plasticity (White, 1947), as shown in Table 23. Also, the dry strength and ease of extrusion are improved. Table 24 shows the green strength of some clay minerals and calcium montmorillonite is high.

Table 23. Water of plasticity (in % by weight)

Kaolinite	8.9–56.03
Illite	17–38.5
Halloysite	33–50
Attapulgite	93
Montmorillonite	83–250

Table 24. Green strength (in kg/cm^2)

Kaolinite	0.34–3.2
Illite	3.2
Calcium montmorillonite	> 5
Halloysite	> 5

Table 25. Linear drying shrinkage of clay minerals (in %)

Kaolinite	3–10
Illite	4–11
Montmorillonite	12–23
Attapulgite	15
Halloysite	5–15

Table 26. Effect of montmorillonite additions on the dry strength of kaolinite

Montmorillonite (%)	Modules of rupture (psi)
0	283
1	391
3	536
5	732

However, a distinct disadvantage is the high shrinkage imparted by montmorillonites (White, 1947), as shown in Table 25. At the same time, there is a dramatic increase in dry strength (White, 1947), as shown in Table 26. Therefore, the ceramic manufacturers must determine what properties need improvement that can be remedied by the addition of a small percentage of montmorillonite. White- and cream-colored bentonites are sometimes added to porcelain enamels as a suspending agent and to lower the firing temperature. If a casting clay is not viscous enough in the mold used to make sanitaryware, a small amount of white- or cream-colored bentonite is added. Very small percentages are also sometimes added to whiteware bodies and electrical porcelains. Also, small amounts of sodium montmorillonite can be added to glazes as a suspending agent (Harman et al., 1944).

6.9. Cosmetics

Sodium bentonite, acid-activated calcium bentonite, hectorite, organoclays, and white bentonite are used in numerous cosmetic formulations. Small quantities, generally of the order of 2% of a formulation, are additives to provide thixotropic and suspension aids, and to improve the structure of liquid systems. It is the combination of swelling, gelling properties, cation exchange capacity, whiteness and brightness, that enable the use of these fine particle flake-shaped bentonites. As thickeners, they are ideal constituents in shampoos and toothpastes. As suspension and dispersion aids, they are especially used in powdered pigments. They can be used as thickeners for a continuous oil phase in skin creams. In pigmented foundation creams, they are used to suspend the pigments and to provide UV protection (Thi Minh Thao et al., 2005). For astringents in gel form in face masks, they are blended with water to form a smooth spreadable paste. Another use of bentonite is to aid in the dispersion of perfume throughout bubble bath formulations. Liquid make-up is essentially pigments dispersed in a viscous base and the biggest problem is to prevent settling of the pigment constituents and this is accomplished by using high viscosity bentonite for thickening. Bentonites are also used in nail lacquers as suspending agents. The use of organoclays in cosmetics will be discussed under the heading organoclays.

6.10. Crayons

White bentonite is used as a filler in pastel-colored crayons and hectorite is used in other crayons as a filler. The addition of the bentonite makes the wax crayons stiffer and less likely to bend at higher temperatures.

6.11. Deodorizers

Calcium bentonite controls odor emitted from cat litter boxes by absorbing the ammoniated compounds which are responsible for the offensive odor. Generally, a clay litter will absorb odors for 3–5 days before the clay in the litter box needs to be replaced. Deodorant additives mixed in the clay litter will double the useable time before the litter needs replacement.

6.12. Dessicants

Calcium montmorillonites that are dried to temperatures high enough to remove most of the interlayer water from between the silicate sheets will avidly absorb water. The temperature of drying is generally between 90 and 150°C in order to not remove all the interlayer water. Total collapse of the layers inhibits the potential to absorb water. Because calcium montmorillonite has two molecular layers of water between the silicate sheets, it has a higher water absorption capacity than sodium montmorillonite which has only one molecular layer. Also, sodium montmorillonite slakes and disintegrates much more readily than calcium montmorillonite. Therefore, calcium montmorillonite is preferred for use as a dessicant.

6.13. Detergents

Sodium bentonite is used as a detergent in dry-cleaning heavily soiled fabrics. The bentonite absorbs the dirt and other staining material and is removed from the fabric with the dry-cleaning fluid. White bentonite is used in detergents after conversion to molecular sieves. This has become a much more prevalent use because the use of phosphate in detergents has been severely restricted in the past several years because of environmental concerns.

6.14. Emulsion Stabilizers

Sodium bentonite is used as emulsifying and stabilizing agents in oil–water systems. Such clays are used both in oil-in-water emulsions and water-in-oil emulsions. An example is the use of bentonite in bituminous-emulsion coatings for the surface protection of concrete. Sodium bentonite is also used in tar and asphalt emulsions.

6.15. Fertilizer

Bentonites are used as additives to chemical fertilizers as diluents to provide the optimum concentration of the needed elements. In arid areas, bentonites with high absorptive capacity and water holding power are used where the soils are coarse and porous and rapidly lose moisture. Liquid fertilizers have become very popular because they are easy to apply and do not create a dust problem. The liquid fertilizer must maintain uniformity in all the handling steps from preparation to application by the farmer. A bentonite must act as a suspension aid and a stabilizing agent in the liquid fertilizer. The best way to determine the effectiveness of the clay to perform the necessary properties to the liquid fertilizer is trial and error because of the chemical complexity of many fertilizers.

6.16. Food Additives

Sodium high swelling montmorillonite is used in wet-mash-type feeds for animals and poultry. An addition of 5% sodium bentonite to 95% dry mash, which with the addition of water, will make a thick wet mash, which will keep the coarse and fine grains in suspension without settling. Chicken and swine feeds are good examples. Bentonite has no food value. It is reported that the addition of about 1% sodium bentonite to the cereal flour mix in bread and other baked cereals reduces staling (Holden, 1948). The addition of a small amount of white bentonite to cake frostings stiffens it so that it does not sag. Also, sodium and calcium montmorillonites, when added to corn and other grains, selectively absorbs and removes alpha-toxins (Kannewischer et al., 2005).

6.17. Fulling Wool

As mentioned previously, calcium montmorillonite, which is termed fuller's earth in Great Britain, was used to absorb the dirt and lanolin from wool in the 1700s and 1800s and perhaps earlier (Robertson, 1986). The process of cleaning the wool was called fulling, thus the name fuller's earth for the absorbent calcium montmorillonite which was used in the process. The process was to use a fulling mill which had feet which struck the wool, water, and fuller's earth mixture to loosen the dirt and allow the clay to absorb the lanolin and dirt particles which were then removed by washing. Each foot struck the mixture 40 blows per minute (Robertson, 1986).

6.18. Herbicides, Insecticides, and Pesticides

The chemical compounds that are used as pesticides, insecticides, or herbicides are highly concentrated so that absorption on a calcium montmorillonite particle (usually in granular form) permits effective distribution and dilution. The pesticide formulations are absorbed on the surface of the granular particle which is incorporated in fertilizers or spread directly on the ground with suitable equipment. In the past and to a limited extent now, the chemical is mixed with pulverized clay and spread as dust or is mixed with water and sprayed as a solution or emulsion directly on the plant or on the ground. In some cases, the clay surface catalyzes the chemical compound and by heating the clay to a temperature of about 600 or 700°C, this problem can be alleviated. It is important that the montmorillonite structure is maintained so the temperature must be below the dehydroxylization temperature.

6.19. Medicines

Robertson (1986) reported that fuller's earth was taken internally for stomach complaints for many centuries. In the First World War, fuller's earth was mixed with food to prevent dysentery. In Germany, Robertson (1986) reported that montmorillonite clay is taken for absorbing poisons, controlling the acidity of the stomach, stomach ache, fermenting and putrifying conditions, and diarrhea. Grim (1962) reported that montmorillonite has been used for a long time in the preparation of pastes, ointments, and lotion for external use. Recently, a hydrothermal bentonite from Nevada has been used to promote joint mobility and flexibility, i.e. arthritis. Testimonials indicate that it is effective (Kriegel, 2004). Bentonite, usually sodium, is taken to relieve stomach ulcers and according to several individuals, it is effective. Bentonite is used as a suspending agent in several medicinal formulations.

6.20. Nanoclays

Nanoclays are ultra-fine clays usually considered to be less than 0.5 μm and commonly less than 0.2 μm. One dimension is in the size range of 1–100 nm. A recent book described *Functional Fillers and Nanoscale Minerals* (Kellar et al., 2003). These ultra-fine clays are very reactive and when incorporated into polymers, ceramics, inks, paints, and plastics, give some exceptional functional properties. Their properties are due to the large surface area to volume ratios. Hectorite and sodium montmorillonite can be exfoliated to single platelets about 1 nm thick, giving high aspect ratios in excess of 100:1 (Schoonheydt, 2002; Harris, 2003). Nanoclays can be incorporated into many thermoplastic polymers, which give improved performance at much lower loadings than required for conventional fillers. Harris (2003) reported that 3–5% nanoclay loadings would compare with 10–50% loadings of a conventional filler. Performance improvements include increased tensile strength, heat deflection temperature, and flame retardance (Fukushima, 2005). The market for nanoclays in flame retardants is estimated to be about 40,000 tons annually. Also, the growth potential for use in automotive composites is very large because of a 7–12% weight reduction on exterior parts. Nano-montmorillonite is used in oxygen-scavenging barrier nylon resins for polyethylene terephthalate bottles used in the hot fill juice market. The market for montmorillonite and hectorite nanoclays will increase significantly in the near future because of many new applications.

6.21. Organoclays

Sodium montmorillonites with a high exchange capacity and hectorite are specially processed to make organoclays. In this process, the exchangeable ions are replaced with organic compounds such as alkylamines and many others (Jordan, 1949). These organoclad montmorillonites are used as thickeners in paints, greases, oil-base drilling fluids, to gel various organic liquids, cleanup oil spills (Carmody et al., 2005), and recently nanocomposites. The nano-montmorillonites are treated with organic molecules which interact with polymers to produce very strong and heat-resistant products. Raussell-Colon and Serratosa (1987) described the mechanisms of interaction and the manner in which organic reactants are arranged on the mineral substrate. The naturally hydrophilic montmorillonites can be changed so that they become organophilic or hydropholic.

6.22. Paint

Montmorillonite clays are used extensively in paints. White bentonites are a preferred material if available. Those which are best are those which carry sodium as the exchangeable cation and are highly colloidal and completely dispersible. In water-based paints, the sodium and/or lithium montmorillonites are suspending and thickening agents. These montmorillonites are also used as an emulsifying agent in both water- and oil-based paint formulations. Organoclays can be tailor made with organic compounds to meet the requirements of different vehicles including lacquers, epoxy resins, and vinyl resins, which are used in paint formulations. These organoclays improve pigment suspension, viscosity, and thixotropy control and are excellent in non-drip emulsion paint.

6.23. Paper

Sodium bentonite is used in the de-inking process to recover cellulose fibers (Murray, 1984). The de-inking process involves heating the recycled paper in a caustic soda solution in order to free the ink pigment. A detergent is then added to release the ink pigment from the cellulose fibers. Sodium bentonite is added to adsorb the ink pigment after which the cellulose fibers are washed to remove the bentonites which carries the ink pigment with it.

Sodium bentonite is also used to prevent agglomeration of pitches, tars, waxes, and resinous material (Murray, 1984). The addition of 0.5% bentonite based on the dry weight of the paper stock prevents agglomeration so that these sizeable globules will not stick to screens, machine wires, press rolls, etc., which causes holes and defects on the paper. Also, there have been claims that the addition of about 2% sodium bentonite at the beater will aid in the retention of filler pigments in the paper stock and also to aid in the distribution of the filler pigments uniformly throughout the paper stock.

In some instances, a small quantity of sodium bentonite has been added to increase the low shear viscosity of certain coating color formulations. Janes and McKenzie (1976) reported that a small addition of sodium bentonite (preferably white in color) to coating formulations improved rheology, smoothness, and opacity.

6.24. Pencil Leads

Pencil leads are comprised of graphite which is bonded with clay. The clay is a mixture of very fine particle size kaolinite and a small amount of bentonite, which improves the plasticity, green strength, and dry strength (Murray, 1961). The clay percentage in a 2H pencil lead is significantly less than the percentage in a 5H pencil lead. The hardness of the lead is controlled by the percentage of the kaolin–bentonite mixture incorporated into the graphite. The mixture of graphite and clay is extruded to form the pencil lead, which is dried and fired to produce the final pencil lead product.

6.25. Pharmaceuticals

Bentonites, particularly sodium bentonite, are used as a suspension aid in many pharmaceuticals. It is also used as a binder in making some pills. Hectorite and white bentonite are preferred for use in pharmaceuticals. The suspending, gelling, and adsorptive properties are valued for use in certain pharmaceuticals.

6.26. Pillared Clays

Pillaring of smectite clay minerals with inorganic cations is an active research area. Pillared clays are processed for use as catalysts (Vaughan and Lussaer, 1980; Figueras, 1988; Turgutbasoglu and Balci, 2005), selective sorbents (Ishii et al., 2005), membranes (Mitchell, 1990), electrochemical and optical devices (Mitchell, 1990), and hosts for enzymes and dyes (Mitchell, 1990). Pillaring is considered to be an ion exchange process and the most prevalent compounds are Al hydrates (Schoonheydt, 1993; Schoonheydt et al., 1994; Dimov et al., 2000). Iron containing pillared clay has been checked in the hydroxylation of phenol (Letaief et al., 2003) which results in higher yields and shorter reaction times. The presence of transition metals in the clay has proven useful to promote different organic reactions (Carrado et al., 1986). Fe containing pillared clays have been studied for the Fisher–Tropsch processes (Bergaya et al., 1991; Rightor et al., 1991). The future of special applications for pillared clays is multitudinous as more and more fundamental and applied research and development is completed.

6.27. Plastics and Rubber

The use of bentonites in plastics was discussed in the section on nanoclays. Bentonite is used in some rubber compounds as an additive to latex for the purpose of thickening and stabilizing (Anonymous, 1937). Hauser (1955) described the use of montmorillonite to set up a thixotropic gel in some latex systems such as in the production of rubber gloves.

6.28. Sealants

The use of high swelling sodium bentonite was discussed in the section on barrier clays. An extensive use is to line irrigation canals and ditches to prevent water from escaping into the adjacent soils where irrigation is not needed. Many farm ponds leak and a possible method to stop the leak or leaks is to spread hay or straw over the pond surface and let it sink to the bottom. Once it has settled, then sodium bentonite is added which sinks and swells to fill the cracks where the water is leaking. The amount of sodium bentonite added is variable, but spreading about a ton per acre will stop most leaks. Another practice is to add a limited amount to the soil next to the foundation and mix it into the soil. The bentonite will swell and prevent water from entering the area adjacent to the foundation. Too much bentonite will cause the foundation to fall inward because of the swelling pressure. Another use is to stabilize what are termed slurry trenches into which concrete is poured. Normally, wooden forms are used, but an alternative and less expensive method is to fill the trench with water and bentonite to form a rather viscous slurry. The bentonite will line the sides of the trench and stabilize the soil or other soft material and the concrete can then displace the slurry to make the form that is required.

6.29. Seed Growth

A relatively new application is to coat seeds with a bentonite slurry which will provide the water to promote the rapid sprouting of the seed when planted. Fertilizer and insecticides can be added to the slurry. This is used primarily in vegetable gardens and greenhouses.

6.30. Tape Joint Compounds

Although palygorskite is the preferred clay for use in tape joint compounds, bentonite is also used. The wallboard joints are filled with an adhesive compound to form a smooth surface for paint or wallpaper. The adhesive film must be very fine and not develop shrinkage cracks. For this reason, non-swelling calcium bentonite is preferred.

6.31. Water Clarification

Wyoming bentonites are used to clarify water because it is easily dispersed and has good adsorptive properties. Dye manufacturers use the sodium bentonite to preferentially adsorb the dye which will sink to the bottom. Bentonite is also used to adsorb paper mill wastes, sewage, and certain industrial wastes (Olin et al., 1942). Heavy metals are removed from wastewaters by Ca and Na bentonites (Alvarez-Ayuso and Garcia-Sanchez, 2003). Cr, Cu, Ni, Zn, and Cd were adsorbed by the bentonites.

REFERENCES

Alvarez-Ayuso, E. and Garcia-Sanchez, A. (2003) Removal of heavy metals from waste waters by natural and Na-exchanged bentonites. *Clay. Clay Miner.*, 51, 475–480.

Anonymous (1937) Use of bentonite in rubber. *Rubber Age*, May, 30–32.

Anonymous (1963) *Foundry Sand Handbook*, 7th Edition. American Foundrymen's Society, Des Plains, IL.

Bergaya, F., et al. (1991) Mixed Al–Fe pillared laponites, preparation, characterization and catalytic properties. Chapter in *Syngas Conversion in Preparation of Catalysts*. Elsevier, Amsterdam, The Netherlands.

Bradley, W.F. (1959) Density of water sorbed on montmorillonite. *Nature*, 183, 1614–1615.

Carmody, O., et al. (2005) Application of organoclays for cleaning up oil spills (Abstract). *Proceedings of the 13th International Clay Conference*, Waseda University, Tokyo, Japan, p. 89.
Carrado, K.A., et al. (1986) Chromium (111) doped pillared clays (PILCs). *Inorg. Chem.*, 25, 4217–4221.
Devaney, F.D. (1956) *Process of Preparing Indurated Pallets of Iron Ore Fines.* US Patent 2,713,172.
Dimov, V.T., et al. (2000) Structural model of Al_{13}-pillard montmorillonite. *Clay. Clay Miner.*, 48, 1–9.
Figueras, F. (1988) Pillared clays as catalysts. *Catal. Rev. Sci. Eng.*, 30, 457–499.
Fukushima, Y. (2005) Organic/inorganic interactions in polymer/clay mineral hybrids, clay. *Science*, 12(Suppl. 9), 79–82.
Griffith, J. (1990) Acid activated bleaching clays. *Ind. Miner. Mag.*, September, 55–67.
Grim, R.E. (1962) *Applied Clay Mineralogy*. McGraw-Hill, New York, 402pp.
Grim, R.E. (1968) *Clay Mineralogy*, 2nd Edition. McGraw-Hill, New York, 596pp.
Grim, R.E. and Johns, W.D. (1957) Compaction studies of molding sands. *Trans. Am. Foundrymen's Assoc.*, 59, 90–95.
Harman, C.G., et al. (1944) Study of the factors involved in glaze-slip control, I–IV. *Am. Ceram. Soc. J.*, 27, 202–220.
Harris, P. (2003) It's a small world-nanominerals growing influence. *Ind. Miner. Mag.*, October, 60–63.
Hauser, E.A. (1955) *Silica Science*. D. Van Nostrand Co., Inc., Princeton, NJ.
Hettinger, W.P. Jr. (1991) Contributions to catalytic cracking in the petroleum industry. *Appl. Clay Sci.*, 5, 445–468.
Holden, E.G. (1948) *Preservation of Baked Cereal Feed.* US Patent 2,443,138.
Ishii, R., et al. (2005) A study of encapsulation and release behaviors of several perfume into a microporous pillared clay mineral and its application to a micro-capsule for a clay/polymer nanocomposite (Abstract). *Proceedings of the 13th International Clay Conference*, Waseda University, Tokyo, Japan, p. 119.
Janes, R.L. and McKenzie, J.D. (1976) The Behavior of Bentonites in Pigmented Paper Coatings—Effects on Rheology, Application, Coated Paper Properties and Printability. Monograph 59, Tappi, Atlanta, GA, pp. 76–80.
Johnson, C.T., et al. (2005) Molecular hydrology—understanding clay–water interactions at the nanoscale (Abstract). *Proceedings of the 13th International Clay Conference*, Waseda University, Tokyo, Japan, p. 54.
Jordan, J.W. (1949) Organophilic bentonite I. *J. Phys. Chem.*, 59, 294–505.
Kannewischer, I., et al. (2005) Characterization of smectites as sorbents of aflatoxin (Abstract). *Proceedings of the 13th International Clay Conference*, Waseda University, Tokyo, Japan, p. 53.
Keith, K.S. and Murray, H.H. (1994) Clay liners and barriers. Chapter in *Industrial Minerals and Rocks*, 6th Edition. Carr, D.D., ed. Society for Mining, Metallurgy and Exploration, Littleton, CO, pp. 435–462.
Kellar, J.J., et al. (2003) *Functional Fillers and Nanoscale Minerals.* Society for Mining, Metallurgy and Exploration, Inc., Littleton, CO, 293pp.

Kendall, T. (1996) Bentonite—major market review. *Ind. Miner. Mag.*, May, 25–37.

Kriegel, C. (2004) Personal communication on SierraSil, Bozeman, MT.

Letaief, S., et al. (2003) Fe-containing pillared clays as catalysts for phenol hydroxylation. *Appl. Clay Sci.*, 22, 263–277.

Low, P.F. (1961) Physical chemistry of clay–water interaction. *Adv. Agron.*, 13, 269–327.

Mielenz, R.C. and King, M.E. (1955) Physical–chemical properties and engineering performance of clays. *Calif. Div. Mine. Bull.*, 169, 196–254.

Mitchell, I.V. (ed.) (1990) *Pillared Layered Structures*. Elsevier, Applied Science, London, 252pp.

Murray, H.H. (1961) *Pencil Clays*. US Patent 2,986,472.

Murray, H.H. (1984) *Kaolins*. Chapter in *Paper Coating Pigments*. Hagemeyer, R.W., ed. Monograph, Tappi Press, Atlanta, GA, pp. 95–191.

Olin, H.L., et al. (1942) Bentonite as a coagulant for sewages and industrial wastes. *Water Works Sewer.*, December, 18–22.

Raussell-Colon, J.A. and Serratosa, J.M. (1987) Reactions of clays with organic substances. Chapter in *Chemistry of Clays and Clay Minerals*. Newman, A.C.D., ed. Mineralogical Society, London, pp. 371–422.

Rightor, F.G., et al. (1991) Iron oxide pillared clay with large gallery height: synthesis and properties as a Fisher–Tropsch Catalyst. *J. Catal.*, 130, 29–40.

Robertson, R.H.S. (1986) *Fuller's Earth—A History of Calcium Montmorillonites*. Mineralogical Society Occasional Publications, Volturna Press, Hythe, Kent, UK, 421pp.

Saeed, A. (1996) Bentonite in animal feed. *Ind. Miner. Mag.*, 346, 47–51.

Schoonheydt, R.A. (1993) The pillaring of clays. Part I: Pillaring with dilute and concentrated Al solutions. *Clay. Clay Miner.*, 41, 598–607.

Schoonheydt, R.A., et al. (1994) The Al pillaring of clays. Part II: Pillaring with $[Al_{13}, O_4(OH)_{24}(H_2O)_{12}]^{7+}$. *Clay. Clay Miner.*, 42, 518–525.

Schoonheydt, R.A. (2002) Smectite-type clay minerals as nanomaterials. *Clay. Clay Miner.*, 50, 411–420.

Thi Minh Thao, H., et al. (2005) Some possibilities of substitution non-natural additives by clay minerals in UV protection creams (Abstract). *Proceedings of the 13th International Clay Conference*, Waseda University, Tokyo, Japan, p. 56.

Turgutbasoglu, F. and Balci, S. (2005) Improvements of catalytic properties of the pillared layered clays and investigation of catalytic activity in CO oxidation (Abstract). *Proceedings of the 13th International Clay Conference*, Waseda University, Tokyo, Japan, p. 55.

Vaughan, D.E.W. and Lussaer, R.J. (1980) Preparation of molecular sieves based on pillared interlayered clays. *Proceedings of the 5th International Conference on Zeolites*. Rees, L.V., ed., Heyden, London, pp. 94–101.

White, W.A. (1947) *The Properties of Clays*. MS Thesis, University of Illinois, IL.

Chapter 7

PALYGORSKITE AND SEPIOLITE APPLICATIONS

The application of palygorskite and sepiolite are as varied as those described for kaolins and bentonites. The elongate shape of these two minerals (Fig. 10) results in unique colloidal properties, especially the resistance to high concentrations of electrolytes. The elongate particles vary in length from about 1 to 10 μm and are approximately 0.01 μm in diameter. This shape and size results in high surface area and high porosity when thermally activated. This elongate needle shape is in contrast to the flake-shaped kaolinite and montmorillonite which leads to some unique applications.

Haden (1963) divided the applications into two broad categories, colloidal and non-colloidal. Colloidal properties result when the particles are dispersed in a liquid medium to the extent that the individual elongate needles are capable of more or less independent motion relative to one another. In the non-colloidal case, the needles are attached to each other to give rigid particles, each of which is comprised of many discrete needles. Table 27 lists some of the important physical and chemical properties of palygorskite and sepiolite.

The internal arrangement of the tetrahedral and octahedral layers of palygorskite and sepiolite is unique in that there are channels through the structure (Fig. 14). These channels are filled with what is termed zeolitic water. When this water is driven off by heating the surface area and thus the sorptivity is increased, chemical compounds that are of the size that

Table 27. Properties of palygorskite and sepiolite

Particle shape	Elongate
Mohs' hardness	2.0–2.5
High surface area	150–320 m^2/g
Moderate base exchange capacity	30–50 meq/100 g
Charge on the lattice	Moderate
API yield	100–115 bbl/ton
Melting point	1550 °C
Sorptivity	High
Water absorption	Up to 100% of the weight of the clay
Oil absorption	Up to 80% of the weight of the clay

Table 28. Applications of palygorskite and sepiolite

Drilling fluids	Floor absorbents
Cat litter	Foundry sand binder
Agricultural carriers	Granulation binders
Tape joint compounds	Laundry washing powders
Paint	Liquid suspension fertilizers
Industrial floor absorbents	Medicines
Adhesives and caulks	Metal drawing lubricants
Animal feed binders	Percolation adsorbents
Anti-caking agents	Pharmaceuticals
Bleaching earths	Polishes
Catalyst supports	Reinforcing fillers
Ceramics	Wax emulsion stabilizer
Cosmetics	

will fit into these channels are readily absorbed. Absorption and adsorption are properties related to surface area. Absorption is the penetration of fluid molecules into the bulk of an absorbing clay, whereas adsorption is the interaction between the fluid molecules and the clay surface.

As previously pointed out, the names palygorskite and attapulgite are used interchangeably in the literature even though the International Nomenclature Committee has determined that palygorskite is the preferred name.

Table 28 lists the many uses of palygorskite and sepiolite (Galan, 1996; Murray, 2005). The first six applications consume the largest tonnages and the remaining uses are listed alphabetically. Each of these uses is discussed. Because of their elongate shape, these minerals are excellent suspension aids in systems with a high electrolyte content, which causes smectites particles to flocculate. Palygorskite and sepiolite particles do not flocculate because of the hindered settling of the elongate crystals.

1. DRILLING FLUIDS

Palygorskite and sepiolite are used as a thixotropic gelling viscosity builder and suspending agent in the drilling of oil and gas wells. Because of their marked stability in the presence of brines and electrolytes (as contrasted with bentonite), these minerals are favored for use. Palygorskite from Senegal and Spain and sepiolite from Spain are used when brines are likely to be encountered in the North Sea, Africa, and the Middle East. Palygorskite from the South Georgia–North Florida area are used in drilling locations in North and South America when brines and salts are encountered. Palygorskite from China is used in China and other East Asian countries and in Australia. The properties needed for drilling mud are as follows from American Petroleum Institute (API, 1962), Specification 13A:

Viscosity	30 cps minimum at 600 rpm
Yield point/plastic viscosity ratio	3 maximum
Filtrate volume	15 cm^3 maximum
Residue	75 μm, 4.0 wt.% maximum

Sometimes the viscosity and mud yield can be improved by adding 1–2% MgO and pugging the mixture. The above measurements are made in water containing 40 g of salt (NaCl) per 100 ml of water. Sepiolite is stable at high temperatures and for this reason, is commonly used in the drilling of geothermal wells. Sepiolite has a mud yield above 150 bbl/ton (Alvarez, 1984) and palygorskite one of 100–125 bbl/ton (Haden and Schwint, 1967).

2. CAT LITTER

Both palygorskite and sepiolite have a high sorptive capacity and therefore make an excellent granular material for use as cat litter. Granular particles usually 16/30 or 20/40 mesh in size absorb the feline waste and contain the offensive odors for several days. Clumping cat litter is made by adding high swelling sodium bentonite to the granules which was described in Chapter 6. Also specific chemical compounds can be added to control odors from the litter so that it does not need to be changed for up to about 10 days.

3. AGRICULTURAL CARRIERS

The high sorptive capacity of palygorskite and sepiolite make these minerals very useful as carriers for pesticides, insecticides, and herbicides. Many of these chemicals are liquids or sticky pastes which would be difficult or impossible to use. Impregnated and absorbed on the granules, the chemicals can be readily applied in the field. Because the granules provide a fairly slow release, the chemical remains active during the germination and initial growth. The particular chemical is mixed with the granules and the treated particles are placed in the ground with the seed.

A good example is a pesticide that kills corn borers which continues to be released as the corn grows, thus protecting the stalk from corn borer damage. Sometimes, the granular surface catalyzes the chemical so that it is ineffective. Heating the granules to a temperature just below the dehydroxylation temperature will often prevent the catalysis. Finely pulverized palygorskite and sepiolite are also used as carriers which after mixing with the chemical can then be dusted or sprayed on the growing plant or on the surface of the ground before the seed germinates and begins to grow. Tests for absorbent granules are made using the General Services Administration's Federal Specification P-A-1056A.

4. TAPE JOINT COMPOUNDS

Finely pulverized palygorskite and sepiolite are used extensively to mix with adhesives used to fill joints and cracks in wall board. The filled joint or crack must be level and smooth and not shrink during drying. The elongate clay particles form a network which does not shrink as the adhesive dries, thus forming a smooth and level surface which can be painted or covered with wallpaper. This is a large and increasing market related to building and home construction.

5. PAINT

Palygorskite and sepiolite are used to replace more costly organic thickeners in emulsion paints, which results in a much more water insensitive film and improved color retention on washing because of the insolubility of the clay thickener. The complex mixture of chemical and pigment compounds that make up a paint system tends to flocculate other minerals used as suspension aids. As mentioned before, the elongate particles provide hindered settling which keeps the paint pigments in suspension. The thixotropic properties of palygorskite and sepiolite reduce sagging and provide easy brushing. Also, these minerals act as emulsion stabilizers serving as protective colloids. Another property is improved flatting for low gloss and matte finish paints.

6. INDUSTRIAL FLOOR ABSORBENTS

Palygorskite and sepiolite granules and pulverized material are extensively marketed as floor sweep compounds. Because of their high sorbent

Table 29. Typical percentage of water and oil absorption based on dry weight of clay

Water absorption	Up to 100% of the weight of the clay
Oil absorption	Up to 80% of the weight of the clay

capacity for both oil and water, the granules and/or dust are spread on water, oil, and grease spills in service stations and factories. The water, oil, or grease is absorbed by the clay granules or dust and can be easily swept up and removed. Table 29 shows the typical percentage of water and oil absorption based on the dry weight of the clay. The lack of inflammability is another positive attribute for these floor sweep compounds.

7. MISCELLANEOUS APPLICATIONS

7.1. Adhesives and Caulks

Palygorskite and sepiolite are used as additives in several types of adhesives and caulks to control viscosity and shrinkage as the adhesives or caulks dry. Also, these elongate minerals provide a controlled gel of uniform composition. A finely pulverized palygorskite or sepiolite is necessary for use in these applications.

7.2. Animal Feed Binders

Both palygorskite and sepiolite are used as binders in making animal feed pellets. The same properties as explained in Chapter 6 for bentonites apply to this application. In addition to binding the feed pellet, these clays are excellent absorbents for aflatoxin. Some preliminary studies also indicated that dioxin is absorbed and is not released in the stomach or intestines of poultry or animals.

7.3. Anti-Caking Agent

Both palygorskite and sepiolite absorb water to about 80% or more of their dry weight as shown in Table 29. Therefore, both of these clays are used as anti-caking agents in dry chemicals which are deliquescent such as ammonium nitrate. Pulverized clay or granules can be used for this application.

7.4. *Asphalt*

Palygorskite is used as an emulsifier in asphalt. A positive property is that it acts as an emulsion stabilizer serving as a protective colloid. The asphalt when emulsified is much easier to apply and mix with aggregates.

7.5. *Barrier Sealants*

As shown by Keith and Murray (2001), a blend of palygorskite and/or sepiolite with sodium bentonite will prevent dessication cracks from forming in a barrier that goes through several wetting and drying cycles. Blends containing about 30–40% palygorskite give a good barrier seal that can be used in landfills and toxic waste dumps. The palygorskite also has an affinity for absorbing heavy metal ions, mercury, and uranium.

7.6. *Bleaching Clays*

Both palygorskite and sepiolite are natural bleaching earths. They are used to clarify automotive oil and many edible oils which are used in cooking. Acid-activated bleaching earth is often more effective but is also more costly. Both of these clays can be used for the selective sorption of organic compounds such as nitrites, ketones, and other polar gaseous hydrocarbons.

7.7. *Catalyst Support and Carrier*

Palygorskite, in particular, is a good catalyst support and carrier. The catalyst is readily available in systems because it is not tightly held on the surface of the palygorskite and thus is released. The high surface area, mechanical strength, and thermal stability are positive attributes in their use as catalyst carriers.

7.8. *Ceramics*

Because of its elongate shape, palygorskite is used in some ceramic formulations to promote high strength both green and dried. Also, small additions to ball clays promote strength and faster water release in sanitaryware and in whiteware improve green strength.

7.9. Cosmetics

Many cosmetic formulations are blends of many diverse chemical compounds and the elongate palygorskite and sepiolite keep these chemicals in suspension and equally dispersed. These clays, because of their high absorbency for water and oil, make excellent face packs to cleanse the skin. They also serve as an adhesive and protective agent because they adhere to the skin and form a protective film. They are used to give the skin opaqueness, eliminate shine, and cover up imperfections.

7.10. Filtration

Because of their elongate shape and the haystack-like mat, these minerals are good filter aids. Needles attach to each other which gives a rigid mat that can be used for filtration of special oils and other liquids. Also, because of their absorbtivity, they are used to dry some oils. It is also used as a percolation adsorbent for the removal of high molecular weight compounds which are absorbed by the large pores. Examples are resins in petroleum oils.

7.11. Foundry Sand Binders

Palygorskite, because of its elongate particle shape, is used when a high temperature high strength bonding clay is needed for special metal castings where bentonite would vitrify and cause serious scabbing on the casting.

7.12. Laundry Washing Powders

Because of their high absorbency, palygorskite and sepiolite are used as additives in laundry washing powders to absorb salts and dirt particles and to keep the soap ingredients uniformly dispersed.

7.13. Medicines

Palygorskite and sepiolite are used in certain liquid medications to keep the compounds in suspension and uniformly distributed. Also, because of their high active surface, drugs such as hydrocortisone can be retained and subsequently released at an appropriate rate (Forteza et al., 1988).

7.14. No Carbon Required Paper

In the past, palygorskite was used as the receptor for ink when the capsules were broken by the typewriter key. The palygorskite was a coating on the paper which received the ink when the capsules were broken on the paper sheet above. This application has almost disappeared because the no carbon required paper now uses other mediums as receptors.

7.15. Oil Refining

Granular palygorskite is used in lube oil percolation towers to decolorize and neutralize the oil. The granules of clay are heat activated at 200–400°C before being placed in the tower. The oil is percolated through the bed of clay until the adsorptive capacity is reduced to the point where the effluent oil reaches a predetermined quality level. The clay can be reused after the adsorbed organic matter is burned off at about 600°C.

Another use is treating jet fuels to remove traces of absorbed moisture and other contaminants. A very finely pulverized palygorskite is used for this application. After the palygorskite is mixed with the jet fuel, it is removed by filtration and discarded.

7.16. Pharmaceuticals

Again, elongate palygorskite and sepiolite minerals are used as suspending agents in complex mixtures of components to prevent settling and separation. Also, these two minerals can act as an adsorbent for toxins, bacteria, and even viruses in the intestine (Martindale, 1982). In addition, they act as a protective coating for the stomach and intestine.

7.17. Polishes

Palygorskite and sepiolite are used in polishes as suspending agents for abrasives to keep them uniformly distributed in the polish. These minerals are relatively soft so do not interfere with the polishing effect of the harder minerals such as corundum, emory, or other polishing compounds.

7.18. Suspension Fertilizers

Liquid fertilizers are used increasingly in the application of plant food. Liquid fertilizers require complete solutions of the components in order to be useful. Typically, about 2% palygorskite is used to stabilize the suspension and prevent settling of any insoluble components. Palygorskite is a good choice for this application because of its highly stable colloidal properties in high concentrations of salts. Suspension fertilizers are fluid mixtures of solid materials suspended in concentrated fertilizer solutions, which are gel-like and stable over extended time periods of 1 or 2 months. These gels can become readily fluid with mild agitation so that the liquid fertilizer can be pumped and uniformly applied to the soil.

7.19. Wax Emulsion Stabilizer

Because both palygorskite and sepiolite are classed as protective colloids and thickeners, they are used to stabilize wax emulsions to prevent separations of the complex mixtures and prevent the breaking of the emulsion.

8. HEALTH EFFECTS OF SEPIOLITE

Santaren and Alvarez (1994) summarized the studies that were conducted to determine the potential health hazards of sepiolite. Because of the health threat caused by asbestos fibers, many other fibrous minerals have been investigated.

Spanish sepiolite was investigated using the methods accepted in assessing the health effects of mineral dusts. The epidemiological investigation of the workforce in the mine and processing plant at the Vallecas–Vicalvaro (Madrid) deposit did not show any association between the inhalation of sepiolite dust and lung disease. In vivo studies including inhalation and intrapleural as well as intraperitoneal inoculations studies have shown that sepiolite is not a health hazard.

REFERENCES

Alvarez, A. (1984). Sepiolite: properties and uses. Chapter in *Palygorskite and Sepiolite Occurrences, Genesis and Uses*. Singer, A. and Galan, E., eds. Developments in Sedimentology Vol. 37. Elsevier, Amsterdam, pp. 253–287.

API (American Petroleum Institute) (1962) *Specification for Oil-Well Drilling Fluid Materials*. API, Dallas, TX.

Forteza, M., et al., (1988). Effects of fibrous clay minerals on dexamethasone stability. *Proceedings of the 10th Conference on Clay Mineralogy and Petrology*, pp. 281–286.

Galan, E. (1996) Properties and applications of palygorskite–sepiolite clays. *Clay Miner*, 31, 443–453.

Haden, W.L. Jr. (1963) Attapulgite: properties and uses. *10th National Conference on Clays and Clay Minerals.* NRC-NAS Monograph No. 12, pp. 284–290.

Haden, W.L. Jr. and Schwint, L.A. (1967) Attapulgite its properties and applications. *Ind. Eng. Chem.*, 59, 58–69.

Keith, K.S. and Murray, H.H. (2001) Sorbent clay minerals and their environmental applications. In *Symposium Proceedings*, ICMAT 2001, White, T. and Sun, D., eds., Vol. 1, pp. 165–171.

Martindale, W. (1982) *The Extra Pharmacopoeia*, 28th Edition. Pharmacological Society of Great Britain, Pharmaceutical Press, London, 2025pp.

Murray, H.H. (2005) Current industrial applications of clays (Abstract). *Proceedings of the 13th International Clay Conference*, Tokyo, Japan, p. 55.

Santaren, J. and Alvarez, A. (1994) Assessment of the health effects of mineral dusts. The sepiolite case. *Ind. Miner. Mag.*, April, 101–117.

Chapter 8

COMMON CLAYS

The term common clays is used by the US Geological Survey and the Society for Mining, Metallurgy, and Exploration for clays, shales, soil clays, and glacial clays that are used primarily for structural clay products. These clays are fine grained and typically exhibit plastic behavior when wet. This plastic clay material can be formed into many desired shapes, dried and fired to produce products with a rock-like hardness. Products that are made from these common clays include structural and face brick, drain tile, vitrified pipe, quarry tile, flue tile, conduit tile, pottery, stoneware, and roofing tile (Murray, 1994).

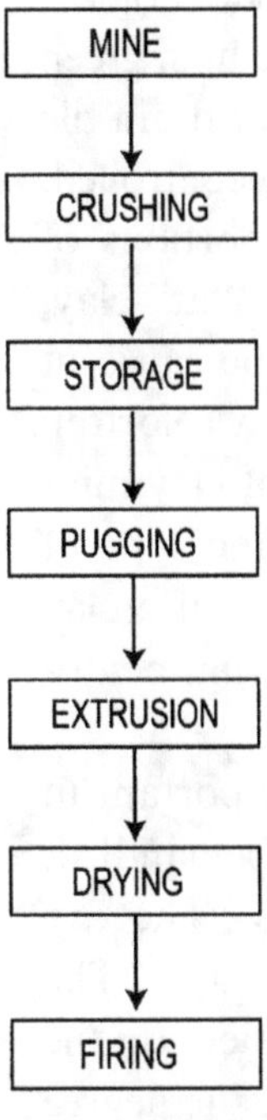

Fig. 64. Process flow sheet for structural clay products.

1. STRUCTURAL CLAY PRODUCTS

These common clays are the most widespread ceramic materials in the world. The products made from these clays do not require elaborate processing. Fig. 64 shows a typical process flow sheet used to make structural clay products. It involves mining from open pits usually located near the processing plant to minimize the transport cost. The clay is stored in sheds usually with open sides to allow some air drying. The clay is crushed, pugged, and extruded into the desired shapes, dried and fired in tunnel or beehive kilns.

Mineralogically, common clays are highly varied and usually the most common clay mineral constituent is illite. Other clay minerals that are frequently present include chlorite, kaolinite, smectite, and mixed-layer clays. Quartz is usually present in almost all common clays. Other non-clay minerals that may be present are feldspar, calcite, dolomite, goethite, hematite, and a small percentage of heavy minerals.

Common clays have a wide range of physical properties thus making them applicable for many different structural clay products. The physical properties that are important are plasticity, green and dry strength, drying and firing shrinkage, vitrification range, fired color and fired strength. The properties desired vary with the type of structural clay product. For example, clay used to make conduit tile must be very plastic, have high green and dry strength, and uniform shrinkage. In the manufacture of drain tile and common brick, these properties do not have to be so closely controlled.

Most clays become plastic when mixed with varying proportions of water. They range from those which are highly plastic, called "fat" clay, to those with low plasticity, which are called "lean" clays. The cause of plasticity has been and is the subject of considerable controversy (Norton, 1948). Particle size and distribution, particle shape, the type of clay mineral present, soluble salts, organic matter, and the amount and type of non-clay minerals are all known to affect plasticity. Sometimes, if a clay has a low plasticity, the addition of a small percentage of sodium bentonite (1–2%) will greatly improve the plasticity.

The green strength and dry strength properties are very important in most types of structural clay products because the ware must be handled. The green strength of a clay is closely related to plasticity. The dry strength is determined by tension, compression, or transverse tests. The transverse test is most commonly employed. Strength is dependent on the proportion of fine particles present, the shape of the particles, the degree of hydration of the colloidal fraction before the sample is prepared, the way in which the test piece is formed, and the extent of drying before

Table 30. Transverse strength of different clays

Type of clay	Modulus of rupture
Washed kaolin	75–200
White sedimentary kaolin, GA-SC	150–166
Ball clay	25–600
Crucible clay	187–691
Refractory bond clay	395–1093
Glass pot clay	173–1068
Sagger clay	46–474
Stoneware clay	94–678
Sewer pipe clay	190–589
Brick clay	50–1500

testing (Holdridge, 1953). The presence of montmorillonite usually increases the dry strength. Table 30 shows a range of transverse strength for different types of clay (Ries, 1927).

Shrinkage, both drying and firing, is another important property of clays used for structural clay products. Drying shrinkage depends on the water content, the type of clay mineral present, and the amount of very fine colloidal material in the clay. The drying shrinkage is high in most very plastic or "fat" clays, which will tend to produce cracking and warping and it is low in sandy or "lean" clays, which will dry to a weak and porous body. The presence of montmorillonite in relatively large amounts (15–25%) causes excessive shrinkage and cracking and slow drying. Shrinkage may be measured in terms of linear dimensions or volume. Firing shrinkage is dependent on the volatiles present, the types of crystalline phase changes, the dehydration characteristics of the clay minerals, and the viscous and surface tension properties (Hyslop, 1953). Firing shrinkage is also measured in terms of linear dimensions or volume.

The temperature range of vitrification or glass formation is another significant property of clays used for structural clay products. Some clays have a very short vitrification range and the temperature in the kiln must be closely regulated. Illites, montmorillonites, and chlorites show evidence of much earlier vitrification during firing than kaolin. Also, the presence of non-clay minerals such as calcite and feldspar may lower the vitrification temperature by acting as fluxes.

Color is an important property of structural clay products. Several factors determine color but iron is usually the primary determinant of color. The color of a structural clay product is influenced by the state of oxidation of iron, the particle size of iron minerals such as hematite and goethite, the firing temperature and degree of vitrification, the proportion of alumina, lime, and magnesia in the clay material, and the composition of the fire-gases during burning (Butterworth, 1953). The best white firing clays contain less than 1% Fe_2O_3. B tan-burning clays contain between 1% and 5% Fe_2O_3, and red-firing clays contain 5% or more Fe_2O_3.

Common clays occur in a variety of environments and in many different rocks across all time periods of the geologic record. The source clay material can be glacial clay, soils, alluvium, loess, shale, weathered and fresh schist, slate, argillate, and underclays or seat earths.

2. LIGHTWEIGHT AGGREGATE

Certain common clays are used to produce lightweight aggregate. The American Society for Testing Materials has published a standard specification governing lightweight aggregates for concrete (Code No. C330-53T, 1955). The unit weight of fine lightweight aggregate cannot exceed $70\,lb/ft^3$ and the unit weight of coarse lightweight aggregate cannot exceed $55\,lb/ft^3$.

Clay and/or shale, which is a common clay, is used as the raw material to make lightweight aggregate. The raw material is crushed and fed into a rotary kiln or sintering machine. The raw material is heated rapidly up to the range between incipient and complete fusion. The bloating and vesiculation require the presence of substances that release gas after fusion has developed a molten jacket around the particles to prevent the escape of the gas. The molten jacket must be viscous enough to prevent the escape of the expanding gas. Conley et al. (1948) and Riley (1951) have investigated the causes of bloating. Several factors are important, most of which are based on chemical and mineralogical composition.

Shales and clays containing illites, chlorite, some montmorillonite, and mixed-layer clays are the most promising sources to make lightweight aggregate. A close relationship exists between the chemical composition and the bloating characteristics of clay and shale. Riley (1951) concluded that the viscosity of the melt produced by firing is determined essentially by the bulk chemical composition based on SiO_2, Al_2O_3, and the total of CaO, MgO, FeO, Fe_2O_3, K_2O, and Na_2O in which optimum viscosity of the molten jacket might be expected. Fig. 19 shows the limits of bloating established by Riley (1951). This was verified by Murray and Smith (1958) in their study of some Indiana shales. Clays and shales of various geologic age and from many formations are used to make lightweight aggregate (Cole and Zetterstrom, 1954; Greaves-Walker et al., 1951; Mason, 1951; Plummer and Hladik, 1951; Burwell, 1954).

REFERENCES

Burwell, A.L. (1954) *Lightweight Aggregate from Certain Oklahoma Shales.* Oklahoma Geological Survey, Mineral Report 24, pp. 1–20.

Butterworth, B. (1953) *The Properties of Clay Building Materials in Ceramics—A Symposium.* Green, A.T. and Stewart, G.H., eds. British Ceramic Society, London, England, pp. 824–877.

Cole, W.A. and Zetterstrom, J.D. (1954) *Investigation of Lightweight Aggregates of North and South Dakota.* US Bureau Mines, Report of Investigation 5065, 1043pp.

Conley, J.E., Wilson, H., and Klinefelter, T.H. (1948) *Production of Lightweight Concrete Aggregates from Clays, Shales, Slates and Other Materials.* US Bureau of Mines, Report of Investigation 4401, 121pp.

Greaves-Walker, A.F., et al. (1951) The development of lightweight aggregate from Florida clays. *Eng. Ind. Exp. Station Bull. Ser.*, 116, 1–24.

Holdridge, D.A. (1953). *The Colloidal and Rheological Properties of Clays in Ceramics—A Symposium.* Green, A.T. and Stewart, G.H., eds. British Ceramic Society, London, England, pp. 60–93.

Hyslop, J.F. (1953) *The Action of Heat on Clays in Ceramics—Symposium.* Green, A.T. and Stewart, G.H., eds. British Ceramic Society, pp. 186–200.

Mason, R.S. (1951) *Lightweight Aggregate Industry in Oregon.* GMI Short Paper 21, 23pp.

Murray, H.H. (1994) Common clay. Chapter in *Industrial Minerals and Rocks*, 6th Edition. Carr, D.D., ed. Society for Mining, Metallurgy and Exploration, Littleton, CO, pp. 247–248.

Murray, H.H. and Smith, J.M. (1958) *Lightweight Aggregate Potentialities of Some Indiana Shales.* Industrial Geological Survey, Report of Progress 12, 42pp.

Norton, F.H. (1948) Fundamental study of clay, VIII, a new theory for the plasticity of clay–water masses. *J. Am. Ceram. Soc.*, 31, 236.

Plummer, N. and Hladik, W.B. (1951) Manufacture of lightweight concrete aggregate from Kansas clays and shales. *State Geol. Surv. Kansas Bull.*, 91, 1–100.

Ries, H. (1927) *Clays, Their Occurrence, Properties and Uses.* John Wiley and Sons, New York, 613pp.

Riley, C.M. (1951) Relation of chemical properties to the bloating of clays. *J. Am. Ceram. Soc.*, 34, 1–20.

Appendix

Appendices A–D describe the test procedures for identifying and evaluation of kaolins, ball clays, bentonite, palygorskite–sepiolite, and common clays. The following tests apply to all the above-mentioned clays:

1. X-ray diffraction to determine the mineral content of the crude and degritted sample.
2. Percent grit (+325 mesh).
3. pH.
4. Particle size distribution of the degritted sample.

The identification of the clay minerals and the non-clay minerals is necessary to determine the amount and type of clay and non-clay minerals present. For example, generally the quartz and micas are concentrated in the grit (+325 mesh) portion of the clay. Also, the grit is generally removed from the clay during processing so that the percent recovery of the portion which is −325 mesh can be calculated.

The pH of the sample gives an indication of the presence of soluble salts, which may be deleterious to the final product. Some kaolins have an alkaline pH which may indicate the presence of calcium and sodium salts which if they cannot be removed by processing will cause high viscosity for paper use and cause a lower temperature of vitrification in ceramic utilization. Also, in the use of kaolins in paint there is a conductivity maximum and the presence of soluble salts may cause a conductivity which exceeds the specification. In ceramic kaolins and ball clays, the presence of montmorillonite can cause excess shrinkage, slow casting rate, and a short and low temperature vitrification range.

In bentonites it is necessary to identify whether or not the clay is a sodium, calcium, or magnesium variety. This can generally be identified by the c-axis d-spacing as sodium montmorillonite has a 12.2 Å spacing and calcium and magnesium montmorillonites have a 14.2–14.8 Å d-spacing.

In kaolins, ball clays, bentonites, and palygorskite–sepiolite clays, a high grit percentage is detrimental. In common clays depending on the type of structural clay product for which it is used, a higher grit percentage may be tolerated.

Appendix A

COMMONLY USED TESTS AND PROCEDURES FOR EVALUATING KAOLIN SAMPLES

Commonly used tests and procedures for evaluating kaolin samples are as follows:

1. Crude clay X-ray diffraction
2. Crude clay moisture
3. Percent grit (+325 mesh) (screen residue test)
4. Degritted X-ray diffraction
5. Crude clay pH
6. Crude clay brightness
7. Degritted clay brightness
8. Crude clay particle size
9. Crude clay Brookfield viscosity
10. Crude clay settling procedure
11. Leaching and brightness test
12. Magnetic separation
13. High shear (Hercules) viscosity
14. Processed clay Brookfield viscosity
15. Processed clay particle size
16. Processed clay brightness
17. Conductivity measurement

1. X-ray diffraction: Pulverize about 2 g of sample to −325 mesh. Press the pulverized clay in the sample holder. Follow operating procedures of the X-ray diffraction unit.

2. Crude clay moisture

   2a. *Apparatus*

   Balance sensitive to 0.1 g
   Ceramic dish
   Drying oven to operate at 105±2°C

   2b. *Procedure*

   Weigh 200 grams (to the nearest 0.1 gram) of crude clay into a ceramic dish of known weight. Place in an oven set at 105°±3°C and allow to dry overnight. Remove from oven and weigh to the nearest 0.1 gram.

   2c. *Calculation*

$$\% \text{ moisture} = \frac{\text{weight loss}}{\text{initial weight of clay}} \times 100$$

3. Percent grit: material coarser than 325 mesh

   3a. *Apparatus*

   325 mesh (44 μm) US standard sieve
   Ceramic dish

Balance sensitive to 0.1 g
Drying oven to operate at 105±2°C
Sodium hexametaphosphate
Soda ash
Small brush
Waring blender

3b. *Procedure*

Dry approximately 200 g of crude clay overnight at 105°C. Mix 5 lb/ton equivalent calgon and 1 lb/ton equivalent soda ash in 500 ml water and mix on the Waring blender. Add 100 g oven dried clay to the mixer cup and mix on low speed for 3 min.Sieve the slurry through a 325 mesh sieve. If a crude particle size is needed, collect screened slurry in a bucket. Rinse screen until all clay has been washed through the sieve. Closely observe residue remaining on sieve. If a significant amount of unblunged clay remains, rinse residue back into Waring container with 500 ml water and mix an additional 3 min. Screen the sample as above. Place screen in oven until dry. Brush residue from screen using small paintbrush into a tared ceramic bowl. Weigh residue to the nearest 0.1 g and record. After sample is degritted, the −325 mesh portion should be X-rayed to determine the mineral content of the degritted sample.

4. Degritted x-ray diffraction: Pulverize about 2 g of sample to −325 mesh. Press the pulverized clay in the sample holder. Follow operating procedures of the X-ray diffraction unit.

5. Crude clay pH test

5a. *Apparatus*

pH meter
Balance sensitive to 0.1 g
250 ml beaker
Deionized water
Spatula

5b. *Procedure*

Weigh 40 g of pulverized clay into a 250 ml beaker. Add 160 ml of deionized water and stir well with a spatula until free of "lumps." Measure pH.

6. Crude clay brightness

6a. *Apparatus*

Brightness meter
Ceramic dish
Drying oven to operate at 105±2°C
Brightness ring and press
Micro-pulverizer
Balance sensitive to 0.1 g

6b. *Procedure*

Grind the sample to a fine powder (< 100 mesh) in a pulverizer. Degritted kaolin should be pulverized for 2 min. Do not use the mortar and pestle. Pack the sample in the brightness meter holder using 32.65 lb of pressure.

7. Degritted clay brightness: Dry the degritted clay and follow the procedure described in item 6.

8. Crude clay particle size

8a. *Apparatus*

Balance sensitive to 0.1 g
Waring blender and container
Deionized water
Plastic 250 ml bottle
250 ml flask
1000 ml beaker

8b. *Procedure*

Using the sample from the crude residue test, determine the percent solids using a 250 ml flask. Determine dry clay content in the 250 ml flask. Dilute percent solids to 7% using a 1000 ml beaker and mix for 3 min in a mixer. Weigh 200 g slurry (14 g dry clay) into a 250 ml plastic bottle labeled with the sample identification. Submit sample for Sedigraph testing.

8c. *Sedigraph procedure*

8c.1. *Sample preparation*

Prepare a 7% by weight solution of clay in water totaling 200 g. (For dry clay use 14 g in 186 ml of water; for slurry clay determine the percent solids and calculate the necessary amount of water to add to reach a 7% solution at 200 g of total weight.) Mix on medium speed for 8 min in Hamilton Beach blender. Agitate for 2–3 min using Lightnin mixer and pour sample into the Sedigraph cup.

8c.2. *Sample analysis*

Follow procedure provided by Sedigraph.

8d. *Apparatus*

Bouyoucos hydrometer
1 l graduated cylinder (2.5 in. diameter) (a larger cylinder may be used, i.e. 1205 ml soil sample cylinder)
Rubber stopper to fit graduated cylinder
Thermometer
Constant temperature water bath (19.4°C) (a constant temperature room may be used if a water bath is not available)
High-speed mixer (Hamilton Beach Model 936)
Table 1: Temperature correction values for hydrometer reading
Table 2: K_N correction coefficients for variation in viscosity of suspending medium
Table 3: Maximum particle size equivalents
Table 4: K_L correction coefficients for a given hydrometer number

8.1. *Particle size test, hydrometer method*

8.1a. *Apparatus*

Bouyoucos hydrometer
1 l graduated cylinder (2.5 in. diameter) (a larger cylinder can be used, i.e. 1205 ml soil sample cylinder)
Rubber stopper to fit graduated cylinder
Thermometer
Constant temperature water bath (19.4°C) (a constant temperature room may be used if a water bath is not available)
High-speed mixer (Hamilton Beach Model 936)
Table 1: Temperature correction values for hydrometer reading
Table 2: K_N correction coefficients for variation in viscosity of suspending medium

Table 3: Maximum particle size equivalents
Table 4: K_L correction coefficients for a given hydrometer number

8.1b. *Procedure*

Particle size distribution is checked by adding 700 ml of deionized water to 50 g of clay (if a larger cylinder is used, the test specimen should be increased accordingly) and agitating it in the Hamilton Beach Model 936 or other high-speed mixer for 8 min on medium speed. Transfer sample to a 1 l graduated cylinder and dilute to the 1000 ml mark. Stopper and shake the cylinder to mix the sample and place it in a controlled temperature bath. Gently immerse the hydrometer into the cylinder, allow it to stabilize, and take reading. Record the temperature as well. This is made easier by using a stopper to suspend the thermometer in place. (If testing more than one sample at a time, be sure to wipe off both the hydrometer and thermometer before placing them into the next sample to avoid contamination.) Read the hydrometer to 0.2 g/l and the temperature to 0.2 °C at 1, 3, 10, 20, 60, 100, 200, 300, and 400 min settling time.

8.1c. *Calculations*

To find the correct grams per liter, add or subtract the appropriate amount shown in Table 1 to or from the actual hydrometer reading. To find the percent in suspension, divide each of the corrected grams per liter by the highest reading. Use the following equation to determine the corrected particle diameter at each sampling time:

$D_t = D_a \times K_L \times K_N$

where:

D_t = particle diameter in microns

D_a = maximum particle diameter in microns from Table 3

K_L = hydrometer correction coefficient from Table 4

K_N = correction coefficient from Table 2

8.1d. *Theory*

To determine particle size with a Bouyoucos hydrometer, use is made of Stokes' law. The maximum diameter of particles in a suspension, based on Stokes' law for assumed conditions, is expressed by the following equation:

$$D_a = \frac{\sqrt{30nL}}{980(G - G')T}$$

where:

D_a = maximum particle diameter in millimeters (equivalent spherical diameter)

n = coefficient of viscosity of suspending medium in poise (in this case, 67°F = 0.0102)

L = distance in centimeters through which the particles settled (32.5 cm)

G = specific gravity of clay particles (2.65 assumed for this formula)

G′ = time period, in minutes, of sedimentation

The above equation gives only the apparent diameter since it assumes that L is a constant. In order to use Stokes' law to determine the diameter of the particle, it is necessary to know the distance through which the particle falls.

The correction coefficient, K_L, for a given hydrometer is computed as follows:

$$K_L = \frac{\sqrt{H_R}}{L}$$

where:

$$H_R = H_l + \frac{1}{2}\left(h - \frac{\text{volume of hydrometer bulb}}{\text{area of graduate}}\right)$$

L = distance in centimeters through which the particles settled (32.5 cm)
Further:
H_l = distance from the top of the bulb to the reading
h = length of the bulb
Variation in viscosity of water at temperatures other than 67°F (19.4°C) is accounted for by the use of the correction coefficient, K_N, and is expressed by the equation:

$$K_N = \sqrt{\frac{n_l}{n}}$$

where:
n_l = viscosity coefficient at a given temperature
n = (0.102) viscosity coefficient of water at 67°F (19.4°C)
Thus making the final equation for the "equivalent spherical particle diameter":
$D_t = D_a \times K_L \times K_N$

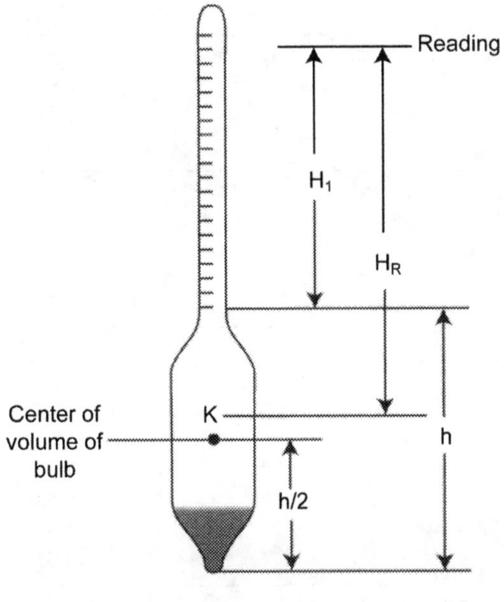

Table 1. Temperature correction value for hydrometer reading

°C	0.0	0.2	0.4	0.6	0.8
16	−1.2	−1.1	−1.1	−1.0	−0.9
17	−0.9	−0.8	−0.7	−0.7	−0.6
18	−0.5	−0.4	−0.4	−0.3	−0.2
19	−0.1	−0.1	0.0	+0.1	+0.1
20	−0.2	+0.3	+0.4	+0.4	+0.5
21	+0.6	+0.6	+0.7	+0.8	+0.9
22	+0.9	+1.0	+1.1	+1.1	+1.2
23	+1.3	+1.4	+1.4	+1.5	+1.6
24	+1.7	+1.7	+1.8	+1.9	+1.9
25	+2.0	+2.1	+2.2	+2.2	+2.3
26	+2.4	+2.4	+2.5	+2.6	+2.7
27	+2.7	+2.8	+2.9	+2.9	+3.0
28	+3.1	+3.2	+3.2	+3.3	+3.4
29	+3.5	+3.5	+3.6	+3.7	+3.7
30	+3.8	+3.9	+4.0	+4.0	+4.1

Table 2. K_N correction coefficients for variation in viscosity of suspending medium

Temperature (°C)	K_N
19.0	1.00
20.0	0.992
21.0	0.980
22.0	0.970
23.0	0.957
24.0	0.946
25.0	0.935
26.0	0.925
27.0	0.915
28.0	0.905
29.0	0.895
30.0	0.885

Table 3. Maximum particle size equivalents

Time (min)	D_A
1	78.42
10	24.80
20	17.54
60	10.12
100	7.84
200	5.53
300	4.53
400	3.92

Table 4. K_L correction coefficients for a given hydrometer number*

Reading	K_L	Reading	K_L	Reading	K_L	Reading	K_L
−5	0.736	11	0.673	27	0.609	43	0.537
−4	0.732	12	0.669	28	0.605	44	0.533
−3	0.728	13	0.665	29	0.601	45	0.528
−2	0.725	14	0.661	30	0.597	46	0.523
−1	0.721	15	0.657	31	0.592	47	0.519
0	0.717	16	0.653	32	0.588	48	0.514
1	0.713	17	0.649	33	0.583	49	0.510
2	0.709	18	0.645	34	0.579	50	0.505
3	0.705	19	0.641	35	0.574	51	0.500
4	0.701	20	0.637	36	0.569	52	0.495
5	0.697	21	0.633	37	0.565	53	0.491
6	0.693	22	0.629	38	0.560	54	0.486
7	0.689	23	0.625	39	0.556	55	0.481
8	0.685	24	0.621	40	0.551	56	0.476
9	0.681	25	0.617	41	0.546	57	0.471
10	0.677	26	0.613	42	0.542	58	0.467
						59	0.462
						60	0.457

*K-values will vary with each hydrometer.

8.1e. *Reference*
Tappi Method T-649

9. Crude clay Brookfield viscosity

9a. *Apparatus*

Brookfield viscometer
Waring blender
Deionized water
Tetrasodium pyrophosphate (TSPP) (0.05 g/ml)
Sodium polyacrylate
Soda ash
600 ml beaker
Thermometer
Balance sensitive to 0.1 mg

9b. *Procedure*

Determine moisture content in the clay (see procedure 2). Weigh out a sample of 500 g equivalent of dry clay. Add enough water for a 70% solids slurry to a Waring blender along with 5#/ton TSPP and 1#/ton soda ash. Start blender and slowly add one-third of total amount of crude clay. If clay is noticeably flocculated, check pH and adjust to 6.5–7.5, targeting 7.0 with the granular soda ash. Begin to add remaining clay and adjust pH. If the slurry continues to flocculate at any point up to and including 70% solids, add 1#/ton increments of sodium polyacrylate up to 6#/ton, while maintaining the 6.5–7.5 pH range. If clay is still flocculated, dilute solids to 67.5%. Blunge 3 min on low speed. Pour sample into a glass 600 ml beaker and adjust temperature to 80±2°F. Measure Brookfield viscosity.

9c. *Brookfield procedure*

9c.1. In Hobart mixer, add 5#/ton equivalent TSPP solution and 1.5#/ton equivalent soda ash in 251 ml total water (221 ml water + 30 ml TSPP solution).

9c.2. Add 600 g clay and mix on speeds 1 and 2 until large lumps of clay disperse and sample becomes fluid.

9c.3. Stop mixing and scrape dry clay from sides of bowl. Replace bowl, cover, and mix on high for 10 min.

9c.4. Check pH and adjust to 6.5–7.5 targeting 7.0. Adjust if necessary using soda ash solution.

9c.5. Pour sample into a container that can be covered or sealed to prevent moisture loss. Check oven dried solids.

9c.6. In a Hamilton Beach mixer cup, weigh 500 g dry clay equivalent and adjust to 70.5% solids.

9c.7. Place on mixer and mix for 5 min. Mix on the highest speed possible being careful as not to lose any clay.

10. Crude clay settling procedure (to classify a sample to a coating clay grade)

10a. *Apparatus*

Balance sensitive to 0.1 g
100 ml graduated cylinder
1000 ml graduated cylinder
10% sodium hexametaphosphate solution (weight/volume)
5% TSPP solution (weight/volume)
10% soda ash solution (weight/volume)
pH meter with electrode
Plastic pan
Large spatula
Cowles mixer
Stainless steel baffled container (13 in. high × 8 in. diameter)

10b. *Procedure*

To 4080 ml of water add 3 lb/ton equivalent of calgon and 2 lb/ton soda ash for coarse soft clays or 5 lb/ton equivalent of TSPP solution and 1 lb/ton soda ash for fine, hard clays. Add 2500 g of crude clay dry clay basis (38% solids). Mix with a Cowles mixer at 4000 rpm for 5 min. Test the pH of the suspension and adjust, if necessary, to 6.5–7.5, targeting 7.0 with 10% soda ash solution. Visually inspect the slurry to insure proper surface tension and that no unblunged clay remains. If clay is not adequately dispersed, add 1equivalent of calgon to the clay and mix on Cowles mixer for 3 min. Repeat this until sample is well dispersed. Pour the sample into a settling pan in which water has been added to dilute the slurry to 32% solids and mix using a spatula. Measure the depth of the suspension in inches. Let slurry settle for 20 min for every inch of depth. Pour the slurry that remains into another container through a 325 mesh screen and measure the volume and percent solids.

11. Leaching and brightness test

11a. *Apparatus*

Technidyne brightness tester (GE brightness)
Pulverizer
Plate glass 8 in. × 8 in.
Cylinders $1\frac{1}{2}$in. diameter and 2tall (stainless steel)
Drill press or plunger of wood or metal with flat, smooth end of a diameter to slip easily into the above cylinder and about 15 cm long

Platform scale, 100 lb minimum capacity (bathroom type)
Drying oven to operate at 80±5°C
pH meter and electrode
Sulfuric acid
325 mesh sieve
Alum (0.65 g/ml solution)
2 l plastic jar
Sodium hydrosulfite (0.125solution)
18.5 cm filter paper: Whatman 201 for fine
Particle clay: Sharkskin for coarse clay
18.5 cm filter bowl with vacuum deionized water

11b. *Procedure*

11b.1. Test the pH of the slurry from the settling procedure and adjust to 2.7–3.2 targeting 3.0 with sulfuric acid.

11b.2. Add 1000 ml of sample into three separate 2 l plastic jars. To one jar, add 9#/ton equivalent sodium hydrosulfite if the crude brightness is greater than 77.0 or 12#/ton equivalent if crude brightness is less than 77.0.

11b.3. Shake the jar containing the sodium hydrosulfite treated sample vigorously for 30 s. Test pH and adjust if necessary to approximately 3.0 with sulfuric acid and allow to stand 10 min.

11b.4. Filter all three jars in separate filter bowls with vacuum using Sharkskin filter paper if clay is soft, Whatman 201 filter paper if clay is hard.

11b.5. When resulting filter cakes are firm, pour 600 ml deionized water onto each filter cake and allow to filter until all additional water is removed.

11b.6. Remove the cakes from the filter bowls. Fold cakes in half with filter paper on outside and place in oven. Dry overnight in oven set at 80±5°C. Remove samples from oven and allow cooling to room temperature.

11b.7. Take approximately 12±2 g of leached and unleached sample and pulverize 30 s in the pulverizer. Crush remaining filter cake using a steel rolling pin until cake has been broken into small pieces and save for viscosity testing.

11b.8. Place a clean glass plate on the platform and two cylinders upon it. Prepare a single test plaque for each sample. Pour the pulverized sample into the cylinders to a depth of approximately 1 in.

11b.9. Depress the plunger, holding both plunger and cylinder to prevent tipping or sliding until the scale shows a weight of 53 lb. Release the pressure and withdraw the plunger.

11b.10. Do not "break" the test plaque from the glass until immediately before testing. Do not rotate the cylinder, as this will "polish" the surface of the plaque and lead to erroneous results.

11b.11. Inspect the surface of the plaque. If surface is broken, rough, or contaminated, discard it and prepare a new one. The surface must not be touched or marred by contact.

11b.12. Carefully place the cylinder containing the plaque on the instrument and measure the brightness with reference to a calibrated working standard in accordance with operating instructions supplied with the instrument.

12. Magnetic separation: Follow procedures set up by the type of magnetic separator unit.

13. High shear (Hercules) viscosity

13a. *Apparatus*

Kaltec Hercules viscometer capable of both 1100 and 4400 rpm testing with 100,000 dyn

Torque sensor, "A" bob, and cup
Balance sensitive to 0.1 g
Hamilton Beach mixer and stainless steel cup
Sodium polyacrylate
Thermometer

13b. *Procedure*
13b.1. Pour sample into cup (sample, cup, and bob should be adjusted to 80±2 °F) to the appropriate level.
13b.2. Lock cup into sleeve and lower bob to stop.
13b.3. Switch viscometer on and determine what rpm is needed in the test, and press the appropriate button (1100 or 4400 rpm).
13b.4. Raise the bob, remove the sample cup, and pour the sample back into the mixing container.
13b.5. Wash and dry the sample cup.

13c. *Minimum Hercules viscosity measurement*
13c.1. Add 2 lb/ton equivalent sodium polyacrylate dispersant. Mix the sample in Hamilton Beach mixer for 2 min. Repeat the viscometer test as described in section 13b. Repeat adding 2 lb/ton increments of sodium polyacrylate and testing the viscosity until minimum Hercules is obtained or until a total of 6#/ton sodium polyacrylate has been added. Report the minimum result as the Hercules viscosity.

14. Processed clay Brookfield viscosity
14a. *Apparatus*
Brookfield viscometer
Hobart mixer with wire whip
600 ml beaker
Balance sensitive to 0.1 g
Drying oven to operate at 105±2 °C
Thermometer
Soda ash
Sodium polyacrylate
pH meter and electrode
Low shear mixer
Hamilton Beach mixer and cup

14b. *Procedure*
14b.1. In Hobart mixer, add 5#/ton equivalent TSPP solution and 1.5#/ton equivalent soda ash in 251 ml total water (221 ml water + 30 ml TSPP solution).
14b.2. Add 600 g clay and mix on speeds 1 and 2 until large lumps of clay disperse and sample becomes fluid.
14b.3. Stop mixing and scrape dry clay from sides of bowl. Replace bowl, cover, and mix on high for 10 min.
14b.4. Check pH and adjust to 6.5–7.5 targeting 7.0. Adjust if necessary using soda ash solution.
14b.5. Pour sample into a container that can be covered or sealed to prevent moisture loss. Check oven dried solids by procedure RS413.
14b.6. In a Hamilton Beach mixer cup, weigh 500 g dry clay equivalent and adjust to 70.5% solids.
14b.7. Place on mixer and mix for 5 min. Mix on the highest speed possible being careful as not to lose any clay.

14c. *Brookfield viscosity measurement*

14c.1. Adjust temperature to 80±2°F.

14c.2. Place sample under viscometer. Use the lowest numbered spindle that can produce a reading of 20–80 on the viscometer scale.

14c.3. Start viscometer rotating at 20 rpm and let rotate for 30 s. Depress clutch and stop viscometer when reading appears in the window.

14c.4. Calculate centipoise and record.

14c.5. Remove spindle and sample. Clean and dry spindle.

14c.6. Add 1#/ton sodium polyacrylate and mix for 1 min in Hamilton Beach mixer set on highest speed possible without losing clay. Repeat steps 14c.1–14c.5 until viscosity minimum is reached. Minimum is signified by an increase in viscosity after the addition of more dispersant.

15. Processed clay particle size

15a. *Sedigraph procedure*

15a.1. *Sample preparation*

Prepare a 7% by weight solution of clay in water totaling 200 g. (For dry clay use 14 g in 186 ml of water; for slurry clay determine the percent solids and calculate the necessary amount of water to add to reach a 7% solution at 200 g of total weight.) Mix on medium speed for 8 min in Hamilton Beach blender. Agitate for 2–3 min using Lightnin mixer and pour sample into the Sedigraph cup.

15a. 2. *Sample analysis*

Follow procedure provided by Sedigraph.

16. Processed clay brightness

16a. *Procedure*

16a.1. Test the pH of the slurry from the settling procedure and adjust to 2.7–3.2 targeting 3.0 with sulfuric acid.

16a.2. Add 1000 ml of sample into three separate 2 l plastic jars. To one jar, add 9#/ton equivalent sodium hydrosulfite if the crude brightness is greater than 77.0 or 12#/ton equivalent if crude brightness is less than 77.0.

16a.3. Shake the jar containing the sodium hydrosulfite treated sample vigorously for 30 s. Test pH and adjust if necessary to approximately 3.0 with sulfuric acid and allow to stand 10 min.

16a.4. Filter all three jars in separate filter bowls with vacuum using Sharkskin filter paper if clay is soft, Whatman 201 filter paper if clay is hard.

16a.5. When resulting filter cakes are firm, pour 600 ml deionized water onto each filter cake and allow to filter until all additional water is removed.

16a.6. Remove the cakes from the filter bowls. Fold cakes in half with filter paper on outside and place in oven. Dry overnight in oven set at 80±5°C. Remove samples from oven and allow cooling to room temperature.

16a.7. Take approximately 12±2 g of leached and unleached sample and pulverize 30 s in the pulverizer. Crush remaining filter cake using a steel rolling pin until cake has been broken into small pieces and save for viscosity testing.

16a.8. Place a clean glass plate on the platform and two cylinders upon it. Prepare a single test plaque for each sample. Pour the pulverized sample into the cylinders to a depth of approximately 1 in.

16a.9. Depress the plunger, holding both plunger and cylinder to prevent tipping or sliding until the scale shows a weight of 53 lb. Release the pressure and withdraw the plunger.

16a.10. Do not "break" the test plaque from the glass until immediately before testing. Do not rotate the cylinder, as this will "polish" the surface of the plaque and lead to erroneous results.

16a.11. Inspect the surface of the plaque. If surface is broken, rough, or contaminated, discard it and prepare a new one. The surface must not be touched or marred by contact.

16a.12. Carefully place the cylinder containing the plaque on the instrument and measure the brightness with reference to a calibrated working standard in accordance with operating instructions supplied with the instrument.

17. Conductivity measurement

This procedure applies to crude, acid dried filter cake, and product clay samples.

17a. *Apparatus*

400 ml glass beaker
Low shear mixer
YSI scientific Model 35 conductance meter
YSI scientific 3418 conductance electrode
Conductivity standard solution
Deionized/demineralized water
Thermometer
Constant temperature bath (optional)

17b. *Procedure*

17b.1. Assure electrode is clean. The bridge inside the electrode should be black. If the metal is seen, the electrode must be platinized using the YSI 34139 Platinizing instrument. Instructions for this can be found on the back of the Platinizing instrument.

17b.2. Adjust the temperature of YSI standard to that stated on standard. Measure conductance. Calculate conductivity by multiplying conductance (micromho (μmho) or millimho (mmho)) by the cell constant (0.1/cm for YSI 3418 electrode). If the conductivity is not within 0.5% of standard value, return to step 17b.1. If the conductivity is within 0.5% of the standard value, proceed to step 17b.3.

17b.3. Disperse 40 g dry clay at 20% solids in deionized/demineralized water.

17b.4. Mix clay 3 min in low shear mixer.

17b.5. Adjust temperature to 25–0.2°C.

17b.6. Place electrode in the sample solution and let rest for 1 min. Record reading.

17b.6.1. Conductivity is calculated by multiplying the measured conductance by the probe cell constant (0.1/cm for the YSI 3418 electrode).

Example: Assume the conductance of a 20% solids kaolin suspension was measured at 8.34 mmho. Multiplying the measured conductance by the probe constant (0.1/cm) yields 0.834 mmho/cm. The result can then be converted to μ μmho/cm.

Appendix B

COMMON TESTS FOR EVALUATION OF BALL CLAY SAMPLES

Procedures for X-ray, pH, screen residue, percent moisture, brightness, particle size distribution, and Brookfield viscosity are the same as those in Appendix A.

1. Specific gravity and percent solids

1a. *Apparatus*

Balance sensitive to 0.1 g
Erlenmeyer flask
Spatula

1b. *Procedure*

1b.1. Record the weight of the Erlenmeyer flask.
1b.2. Pour samples into the flask until the sample is slightly above the rim of the flask. If any spillage occurs all excess must be removed from the flask.
1b.3. Using a spatula, scrape off all sample which is above the rim of the flask.
1b.4. Record weight of the flask and sample.

1c. *Calculations*

$$\text{Specific gravity} = \frac{(\text{weight of flask} + \text{sample}) - (\text{weight of flask})}{\text{Volume of flask}}$$

$$\%\ \text{solids} = \frac{\text{specific gravity} - 1}{\text{specific gravity}} \times 1.625 \times 100$$

Example:

Weight of flask = 208
Weight of flask and sample = 1167
Volume of flask = 560 ml

$$\text{Specific gravity} = \frac{1167 - 208}{560} = 1.712$$

$$\%\ \text{solids} = \frac{1.712 - 1}{1.712} \times 1.625 \times 100 = 67.6\%$$

2. Sulfate and percent transmission

Purpose: To determine the sulfate content of a clay slurry and the percent transmission of the filtrate from the slurry.

2a. *Apparatus*

Visible spectrophotometer
Laboratory mixer
Baroid filter press
Whatman #42 filter paper

2b. *Sample preparation*

2b.1. Clay: The equivalent of 300 g of dried clay is added to 510 g of distilled water. The slurry is mixed until all of the clay is dispersed.
Slurry or slip: The sample is agitated for 1–2 min at low speed.

2c. *Procedure*

2c.1. *Percent transmission*

2c.1a. The wavelength on the spectrophotometer is set at 400 nm.
2c.1b. Distilled water is inserted into the test tube which is then placed into the spectrophotometer.
2c.1c. The spectrophotometer is calibrated to 100%.
2c.1d. Filtrate is inserted into the test tube which is then placed into the spectrophotometer.
2c.1e. The percent transmission is read directly.

2c.2. *Sulfate*

2c.2a. A solution with the following amounts is prepared.
2 ml (or more according to the filtrate).
48 ml (or amount to equal 50 ml) of distilled water.
2c.2b. Mix in 50 ml flask.
2c.2c. Pour solution into 100 ml beaker and add 5 ml of solution of one-third salt acid reagent and two-thirds glycerine.
2c.2d. Set wavelength on the spectrophotometer to 500 nm.
2c.2e. Pour solution into test tube and insert into spectrophotometer.
2c.2f. Calibrate spectrophotometer to 100%.
2c.2g. Add one scoop of $BaCl_2$ crystal and mix with the magnetic stirrer for 3 min.
2c.2h. Pour solution into test tube and insert into spectrophotometer.
2c.2i. Record percent transmission.

2d. *Calculation*

2d.1. *Sulfate*

$$\text{Dry basis} = \frac{(\text{constant from chart})(H_2O/\text{clay ratio})(\text{volume of flask})}{\text{amount of filtrate used}}$$

$$\text{Wet basis} = \frac{(\text{constant from chart})(\text{volume of flask})}{\text{amount of filtrate used}}$$

a. Constant from chart is determined from percent transmission.
b. H_2O/clay ratio is the percent water/percent dry material from original slurry or slip.

3. Casting rate and percent moisture retention

Purpose: To determine the ability of a slurry to form a solid cast and the amount of water remaining in the cast.

3a. *Apparatus*

Laboratory mixer
Drying oven
Brookfield viscometer
Baroid filter press
Whatman #42 filter paper
Balance sensitive to 0.1 g
Dessicator

3b. *Sample preparation*

3b.1. Approximately 700 g of sample is weighed in a 1000 ml beaker.
3b.2. Mix sample for approximately 1 min.
3b.3. Transfer sample to beaker and measure the Brookfield viscosity.
3b.4. Record reading.
3b.5. Return sample to 1000 ml beaker.
3b.6. Add two drops of deflocculant.
3b.7. Repeat steps 2–6 until minimum viscosity is reached.

3c. *Procedure*

3c.1. Measure out 200 ml of sample.
3c.2. Filter 200 ml of sample through a baroid filter press using Whatman #42 filter paper with a pressure of 90 psi (filter slurries for 1 h).
3c.3. After filtration period, remove baroid and allow sample to drain for 5 min.
3c.4. Remove cast from baroid and record weight. This is the casting rate.
3c.5. Place cast into drier for a minimum of 18 h.
3c.6. Remove cast from drier and place in dessicator until it reaches room temperature.
3c.7. Record weight of cast.

3d. *Calculations*

Casting rate: Read directly from scale

$$\%\text{ moisture retention} = \frac{\text{wet weight of cast} - \text{dry weight of cast}}{\text{dry weight of cast}} \times 100$$

4. Modulus of rupture (MOR)

Purpose: To determine the dry strength of a clay.

4a. *Apparatus*

Plaster mold which forms bars 3/4 in. × 7 in.
Low shear mixer
Drying oven
Dessicator
Compression tester
Calipers

4b. *Sample preparation*

4b.1. Clay and slurry.

4b.1a. Clay and water are mixed together to form a slurry with a specific gravity of 1.519 (55.52% solids).
4b.1b. Water is added to a slurry so that a specific gravity of 1.519 is achieved.

4b.2. Mix in the weight of quartz which will give a 50% clay and 50% flint body.

4c. *Procedure*

4c.1. Add deflocculant to sample until minimum Brookfield viscosity is reached.
4c.2. Cast samples in plaster molds forming bars 3/4 in. diameter and 7 in. long. Allow bars to cast overnight.
4c.3. Remove bars from mold and allow to air dry for 6 h.
4c.4. Place bars in drier at 240°F overnight.
4c.5. Place bars in dessicator and allow to cool to room temperature.
4c.6. Remove bars one at a time from dessicator and break on a compression tester with a span of 2 in.
4c.7. Record breakage pressure in pounds and accurately measure the diameter of the bar at the breaking point in inches with calipers.

4d. *Calculation*

$MOR = 8PL/3.1416d^3$

where:

P = span (2 in.)

L = load (lb)

d = diameter (in.)

Average at least five readings and discard any values exceeding ±10% of the average.

4e. *Example*

$P = 2$ in.

$L = 50$ lb

$D = 0.89$ in.

$MOR = 8(2)(50)/3.1416(0.89)^3$

MOR = 361 psi

5. Dry and fired shrinkage

Purpose: To determine shrinkage during the drying and firing cycles.

5a. *Apparatus*

Drying oven

Furnace that will reach 1236°C or 2257°F

Caliper

Plaster mold for 180 mm × 19 mm round bars

5b. *Procedure*

5b.1. A slip is reduced to minimum viscosity by addition of sodium polyacrylate or sodium silicate.

5b.2. The slip is then cast into plaster molds, forming 180 mm × 19 mm round bars.

5b.3. The bars are allowed to cast overnight.

5b.4. Upon removal from the mold a caliper is used to etch marks 100 mm apart which is used for the wet length (WL).

5b.5. The bars are air dried for at least 6 h and dried at 80°C overnight. The dry length (DL) of the bars is measured and the dry shrinkage calculated.

5b.6. The bars are then fired to cone 8 in an electric furnace, after which time the fired length (FL) is measured and the fired and total shrinkage are calculated.

5c. *Calculations*

%DS = percent dry shrinkage

%FS = percent fired shrinkage

%TS = percent total shrinkage

WL = wet length

DL = dry length

FL = fired length

$$\%DS = \frac{WL - DL}{WL} \times 100$$

$$\%FS = \frac{DL - FL}{DL} \times 100$$

$$\%TS = \frac{WL - FL}{WL} \times 100$$

5d. *Example*

WL = 100 mm

DL = 98.2 mm

FL = 96.1 mm

Calculated: %DS, %FS, %TS

$$\%DS = \frac{100 - 98.2}{100} \times 100 = 1.8$$

$$\%FS = \frac{98.2 - 96.1}{100} \times 100 = 2.14$$

$$\%TS = \frac{100 - 96.1}{100} \times 100 = 3.9$$

6. Deflocculation demand

Purpose: To determine the amount of deflocculant required to reduce the viscosity of a clay to minimum.

6a. *Apparatus*

Laboratory drying oven
Syringe
Variable speed mixer
Brookfield viscometer
80 mesh screen
500 ml container

6b. *Sample preparation*

6b.1. The clay sample is dried overnight to 0% moisture.
6b.2. A slurry with a specific gravity of 1.30 is produced by adding 300 g of clay to 510 g of water. The slurry is mixed until all clay is dispersed.
6b.3. The slurry is screened through an 80 mesh screen.
6b.4. A syringe is filled with a solution of 50% deflocculant and 50% water.
6b.5. The syringe is weighed, this is the initial weight of deflocculant.
6b.6. If slurry is too thick (cannot get reading) deflocculant should be added until a reading is achievable. The syringe should again be weighed with this weight being the initial weight of deflocculant.

6c. *Procedure*

6c.1. Mix sample for approximately 1 min.
6c.2. Transfer slurry to beaker and measure the Brookfield viscosity.
6c.3. Record reading.
6c.4. Add an increment of deflocculant. For high sulfate clays add five drops and for easily deflocculated clays add three drops.
6c.5. Repeat steps 1–3.
6c.6. Steps 1–4 are repeated until minimum viscosity is reached.
6c.7. Reweigh syringe and record as final weight.

6d. *Calculations*

6d.1. Amount of deflocculant required

$$\text{Total weight deflocculant} = \frac{\text{initial weight of syringe} - \text{final weight of syringe}}{2}$$

6d.2. Percentage of deflocculant required

$$\%\ \text{deflocculant} = \frac{\text{total weight deflocculant}}{\text{dry weight of clay}} \times 100$$

6d.3. Percent deflocculant per increment (used to develop deflocculation curve)

$$\%\ \text{deflocculant per increment} = \frac{\text{percent deflocculant}}{\text{total number of increments}}$$

*To develop deflocculation curve the y-axis is the viscosity axis (360, the viscosity recorded at the increment number) and the x-axis is the percent deflocculant axis (the increment number × percent deflocculant per increment)

6e. *Example*

Initial syringe weight = 20.000
Final syringe weight = 16.000

Increment number	Viscosity
1	280
2	300
3	312
4	312

$$\text{Total weight of deflocculant} = \frac{20.000 - 16.000}{2} = 2.000$$

$$\%\ \text{deflocculant} = \frac{2000}{300} \times 100 = 0.667$$

$$\%\ \text{deflocculant per increment} = \frac{0.667}{4} = 0.167$$

*Deflocculation curve

Increment number	Viscosity	% deflocculant
1	280	0.167
2	300	0.334
3	312	0.501
4	312	0.667

7. Percent total organic content

Purpose: To determine the percent organic material in the clay.

7a. *Apparatus*

Laboratory drying oven
Hot plate
Evaporator dish
Balance sensitive to 0.1 g
Pulverizer
Hydrogen peroxide

7b. *Sample preparation*

7b.1. Sample is dried in laboratory drier. Heat in drier must be less than 140 °F to prevent loss of volatile organics.

7b.2. Sample is pulverized.

7b.3. Sample is placed in drier for 1 h.
7b.4. Remove sample from drier and place in dessicator. Allow sample to come to room temperature.

7c. *Procedure*
7c.1. Record weight of evaporator dish.
7c.2. Weigh approximately 2 g of sample onto evaporator dish.
7c.3. Record weight of sample and evaporator dish.
7c.4. Add 25 ml of distilled water and 5 ml of peroxide.
7c.5. Place on top of hot plate which is set at 60–70°C.
7c.6. As the solution warms bubble will appear. When the bubbles dissipate add another 5 ml of peroxide.
7c.7. When bubbling dissipates a second time, place evaporator dish in a drier set at 105 °F until the sample is dried.
7c.8. Place evaporator dish in dessicator and allow to reach room temperature.
7c.9. Record weight of evaporator dish and sample.

7d. *Calculations*
W1 = original sample weight = step 3 – step 1
W2 = final sample weight = step 9 – step 1

$$\% \text{ total organic} = \frac{W1 - W2}{W1} \times 100$$

7e. *Example*
Weight of evaporator dish = 21.000 (step 1)
Weight of evaporator dish and sample = 23.000 (step 3)
Weight of evaporator dish and sample = 22.750 (step 9)
W1 = 23.000–21.000 = 2.000
W2 = 22.750–21.000 = 1.75

$$\% \text{ total organic} = \frac{2.000 - 1.750}{2.000} \times 100 = 12.5$$

8. Brightness and color of fired disks
Purpose: To determine the brightness and color of a fired clay.
8a. *Apparatus*
Drying oven
Balance sensitive to 0.1 g
Laboratory press and pressing die (1 in. in diameter)
Colorimeter (Hunter)
Dessicator
Laboratory pulverizer
Firing oven

8b. *Sample preparation*
8b.1. The sample is dried in a laboratory drier.
8b.2. The sample is removed and ground to a fine powder.
8b.3. Sample is placed back in the drier for 1 h.
8b.4. Remove sample from drier and place in dessicator until it reaches room temperature.
8b.5. Weigh out 10 g of sample.
8b.6. Place sample into pressing die.
8b.7. Place pressing die into laboratory press and apply 12,000 lb force using the press.

8b.8. Remove disk from the pressing die.
8b.9. Fire in a laboratory kiln to 1236°C at a ramp rate of 3°C/min and a 30 min soak time.
8b.10. Calibrate colorimeter as described in the "Operator's Manual."
8b.11. Place disk in colorimeter.

8c. *Procedure*

Brightness:

1. Tappi Filter is placed in the "In" position.
2. Press "Brightness" and "Read."
3. Record $L \times a \times b$ numbers.

All readings are read directly from the colorimeter.

Appendix C

COMMONLY USED TESTS TO EVALUATE BENTONITE SAMPLES

DRILLING FLUID PRODUCTS

American Petroleum Institute (1997) *Standard Procedure for Field Testing Water-Based Drilling Fluids*, API Recommended Practice 13B-1, 2nd Edition. Washington, DC.

Items Covered

- Mud weight
- Viscosity and gel strength
- Filtration
- Sand
- Methylene blue capacity
- Pit
- Shear strength measurement
- Calibration of glassware, thermometers, viscometers, and mud
- Balances

FOUNDRY PRODUCTS

American Foundrymen's Society (AFS) (1983) *Mold and Core Test Handbook*, 2nd Edition. AFS, Inc., Des Plaines, IL, 457pp.

Items Covered

- Green compression strength
- Dry and baked compressive strength
- Green shear strength
- Green tensile strength
- Dry shear strength
- Methylene blue test
- Moisture determination
- Compactability of molding sand mixtures
- Preparation of standard permeability test specimen
- Mold permeability test
- Moldability of molding sand mixtures

- Hot compressive strength
- Soluble calcium in bentonite

IRON ORE PELLETIZING

ASTM E946 (1996) Standard Test Method for Water Absorption of Bentonite Porous Plate Method, in Annual Book of ASTM Standards.

ENVIRONMENTAL SEALING

ASTM D5890 (1999) Standard Test Method for Swell Index of Clay Mineral Component of Geosynthetic Clay Liners, in Annual Book of ASTM Standards.

SOME BENTONITE TESTING METHODS

A. Wet screen analysis
B. Viscosity (Brookfield)
C. Filtration
D. Absorption capacity: water and oil
E. Percent expandability

Wet Screen Analysis: Bentonite

1. Prepare a 3–5% solids slurry. Record the weight of the dry clay.
2. Blend the slurry for about 5 min or until all lumps are dispersed.
3. Pour the slurry through a 100 mesh screen and thoroughly rinse the material remaining on the screen to remove all clay.
4. Weigh a clean drying dish and record the weight.
5. Transfer the material left on the screen to the drying dish, being careful not to lose any material.
6. Place the dish in a low temperature (100°C) oven to dry.
7. Weigh the dish and the dried grit and calculate the grit weight: total weight−dish weight = grit weight.
8. Calculate the grit percent: grit weight/sample weight = grit percent.

Appendix D

PALYGORSKITE–SEPIOLITE LABORATORY TESTS

1.Test: API (American Petroleum Institute) yield value (API Specification 13A)

1a. *Purpose*: This test is used for determining palygorskite and sepiolite clay gelling properties in saturated salt water as a performance function of drilling fluid for the petroleum industry.

1b. *Equipment and materials*

FANN model 35 motor driven direct indicating viscometer.
FANN viscometer cup.
Hamilton Beach stainless steel mixer cup: No. M110D or equivalent.
Timer: precision of 0.1 min.
Balance sensitive to 0.01 g.
Graduated cylinder: $500 \pm cm^3$ (ml).
Defoamer (octanol-1 can be used for this purpose).
Saturated salt solution (NaCl in distilled water).

1c. *Procedure*

1c.1. Place 350 ml of saturated salt solution into the Hamilton Beach cup.

1c.2. Weigh out 20.0 g of the clay to be tested. While stirring, add the clay slowly to the cup to avoid clumping or caking on the sides of the cup.

1c.3. Immediately after all the clay has been added, set timer and stir for twenty (20) minutes.

1c.4. Check the solution visually after five (5) minutes. If necessary, remove the container from the mixer and scrape its sides with a spatula to dislodge any clay adhering to the container walls.

1c.5. Ensure that all material clinging to the spatula is incorporated into the suspension. Be sure to stop timer during this period of time.

1c.6. Transfer the solution into the viscometer cup. Add two to three drops of defoamer as necessary to break the surface froth. Place the viscometer cup on the FANN viscometer and record the dial reading at 600 rpm (when constant value reached). Reading temperature shall be at 77 ± 2°F (25 ± 1°C).

1c.7. Determine the yield in barrels/ton (bbl/ton) from Table 1.

1d. *Notes*

1d.1. The saturated salt solution is prepared by adding excess sodium chloride (NaCl) to distilled water and decanting.

1d.2. Table 1 is based on a one-point, constant-slope technique for determining the yield of barrels of fifteen (15) centipoise mud that can be obtained from one (1) ton of clay.

Table 1. Yield in bbl/ton

FANN reading	bbl/ton	FANN reading	bbl/ton
5.0	44.0	65.0	155.0
10.0	61.0	66.0	156.0
15.0	73.0	67.0	157.0
16.0	75.0	68.0	159.0
17.0	78.0	69.0	160.0
18.0	80.0	70.0	162.0
19.0	82.0	71.0	163.0
20.0	84.0	72.0	164.0
21.0	86.0	73.0	165.0
22.0	88.0	74.0	167.0
23.0	90.0	75.0	168.0
24.0	92.0	76.0	169.0
25.0	94.0	77.0	170.0
26.0	95.0	78.0	172.0
27.0	97.0	79.0	173.0
28.0	99.0	80.0	174.0
29.0	100.0	81.0	176.0
30.0	102.0	82.0	177.0
31.0	104.0	83.0	178.0
32.0	106.0	84.0	179.0
33.0	108.0	85.0	181.0
34.0	110.0	86.0	182.0
35.0	111.0	87.0	183.0
36.0	113.0	88.0	184.0
37.0	115.0	89.0	185.0
38.0	116.0	90.0	187.0
39.0	118.0	91.0	188.0
40.0	120.0	92.0	189.0
41.0	121.0	93.0	190.0
42.0	123.0	94.0	191.0
43.0	124.0	95.0	193.0
44.0	126.0	96.0	194.0
45.0	127.0	97.0	195.0
46.0	129.0	98.0	196.0
47.0	130.0	99.0	197.0
48.0	132.0	100.0	198.0
49.0	133.0	101.0	200.0
50.0	134.0	102.0	201.0
51.0	136.0	103.0	202.0
52.0	137.0	104.0	203.0
53.0	139.0	105.0	204.0
54.0	140.0	106.0	205.0
55.0	142.0	107.0	206.0
56.0	143.0	108.0	207.0
57.0	144.0	109.0	208.0
58.0	145.0	110.0	209.0
59.0	147.0	111.0	210.0
60.0	148.0	112.0	212.0
61.0	150.0	113.0	213.0
62.0	151.0	114.0	214.0
63.0	152.0	115.0	215.0
64.0	153.0		

2. Test: Dispersion viscosity

2a. *Purpose*: This test is used as a measure of the viscosity properties of a gelling clay in distilled water without additives.

2b. *Equipment and materials*

Beaker: 600 ml standard form
Brookfield viscometer model RVT with a number four (#4) spindle
Timer
Balance sensitive to 0.1 g
Waring blender: model #31BL92
Distilled water

2c. *Procedure*

2c.1. Tare the Waring blender container on the balance scale. Add seven hundred forty-four (744) grams of distilled water.

2c.2. Separately, weigh out fifty-six (56) grams of the clay to be tested on the balance scale. Add this material to the blender container.

2c.3. Turn on the Waring blender with the rheostat initially set at forty (40) percent. Slowly increase rheostat to one hundred (100) percent (120 V) to ensure that the solution is not splashed from the container during initial agitation. At 100% rheostat setting (120 V), set timer and blend at low speed for exactly five (5) minutes.

2c.4. Transfer the gelled solution to a 600 ml standard form beaker.

2c.5. Determine the Brookfield viscosity at twenty (20) rpm with a number four (#4) spindle. Take one (1) reading after twenty (20) seconds.

2c.6. Multiply the reading by 100 to obtain viscosity in centipoise.

2d. *Notes*

2d.1. It is important that the reading be taken at exactly twenty (20) seconds. The viscosity of the dispersion will decrease with time as the spindle shears the solution. An early or late reading will create inconsistent results when duplicating the test.

2d.2. The test is sensitive with crude clay using distilled water. Normally, viscosity results will be in excess of 2500 centipoise. If viscosity is very low (0–1000 centipoise) with crude ore, run a standard pH, add 1.0% MgO, and retest. Discuss results with Laboratory Manager or Chief Technician.

3. Test: Thixotropic index

3a. *Purpose*: This test is used to compare the ratio of the apparent viscosity at one shear rate to the apparent viscosity at a second shear rate for a given formulation at a prescribed temperature.

3b. *Equipment and materials*

Clay gelling suspension: per any suspension formulation designed to test with Brookfield viscometer.
Brookfield viscometer: model and spindle per suspension test requirement.

3c. *Procedure*

3c.1. Any clay suspension which has been made up to test viscosity using the Brookfield viscometer may be used.

3c.2. The viscosity can be determined at six (6) rpm and at sixty (60) rpm and the thixotropic index (TI) calculated:

$$\frac{\text{Centipoise at 6 rpm}}{\text{Centipoise at 60 rpm}} = \text{TI}$$

3c.3. Typical results:

	A	B	C	Oil
6 rpm	5000	6000	6500	500
60 rpm	1200	1400	1600	500
TI	4.2	4.3	4.7	1.0

Note: A, B, C are different samples to illustrate thixotropic index.

As the value of the TI increases, the more "thixotropic" the composition is. If the TI is less than 1.0, the formulation is dilatent.

3d. *Notes*

3d.1. For water, syrup, silicone fluids, and many oils, the TI will normally be 1.0 (or very close to 1.0) since they are Newtonian fluids. However, for clay stabilized suspension fertilizers, the TI is never equal to 1.0 because they are pseudo-plastic and exhibit shear thinning.

3d.2. Temperature control is important in this determination. Pick a temperature that is easy to use and stick with it. The age of the formulation is also important. TI values generally drop on aging.

4. Test: Tyler Ro-Tap screen analysis

4a. *Purpose*: This test is used to determine the particle size distribution of granulated products. An appropriate sample is placed on the top sieve of a stacked set of US standard sieves and shaken for a specified period of time. The weight percent of the granules retained on each sieve and bottom pan determines the distribution.

4b. *Equipment and materials*

W.S. Tyler Ro-Tap testing sieve shaker or equivalent.
Sieves: US standard sieves or equivalent. Sieve dimensions are one (1) inch (26 mm) in height and eight (8) inches (203 mm) in diameter. Sieve stacks for different products shall be as outlined in Table 2.
Bottom received pan and top sieve cover.
Hard bristle or soft brass brush.
Balance sensitive to 0.01 g.
Timer.

4c. *Procedure*

4c.1. Pour the sample to be tested through a one-half (1/2) inch splitter and repeat five (5) times to thoroughly mix the sample.

4c.2. Split out a representative quantity of the sample as near to 100 g as possible. Weigh to the nearest gram to ensure weight is reasonably close.

4c.3. Stack the sieves per Table 2 product mesh specifications for the product to be tested.

4c.4. Transfer the representative sample onto the top sieve, cover, and place the sieve stack on the Ro-Tap shaker. Be sure the latch is in place. Set the automatic timer for eight (8) minutes and energize the Ro-Tap.

4c.5. When shaking is complete, remove the sieve stack from the Ro-Tap. Quantitatively transfer the granular material on the top sieve to a tared weighing pan, using a stiff bristle brush, and weigh to the nearest 0.1 g. Record the weight. Repeat this procedure for material retained on each sieve and the bottom receiver pan. The tare weight is accumulative. The total weight will appear as a negative when the accumulated sample is removed from the tared weighing pan.

4c.6. To calculate the particle size distribution of each mesh fraction, divide the weight obtained from each sieve by the total weight.
Multiply by 100 to obtain the percentage on each sieve:

$$\% \text{ mesh fraction} = \frac{\text{Sieve fraction weight}}{\text{Total weight}} \times 100$$

Table 2. Sieve stacks

8/16	16/30	20/40	24/48 and 30/60	6/30	60/100
4	12	12	12	6	40
6	16	16	20	8	50
8	18	18	25	10	60
12	20	20	30	16	70
16	25	25	40	20	80
18	30	30	50	30	100
20	40	40	60	60	Pan
60	50	60	70	100	
Pan	60	100	80	Pan	
	Pan	Pan	100		
			Pan		

5. Test: Wet screen analysis

5a. *Purpose*: This test is used to determine the undispersible particles present in finely ground clay products.

5b. *Equipment and materials*

Screen: eight (8) inch, #325 US standard testing sieve.
Three (3) inch, #325 US standard testing sieve (deep frame wet test sieve).
Sterling multimixer, model 9B with 9B29X impeller blades or equivalent.
Multimixer cup to fit model 9B mixer.
Dispersant: certified grade tetrasodium pyrophosphate (TSPP).
Evaporating dishes: #3 Coors flat bottom dishes or equivalent.
Oven maintainable at 220 ± 5 °F.
Dessicator.
Balance sensitive to 0.01 g.
Wash bottle.

5c. *Procedure*

5c.1. Make up a 3.0% solution of TSPP as follow.
5c.2. Add 800 ml of distilled water to a clean, dry one (1) liter beaker.
5c.3. Weigh out exactly twenty-four (24) grams of TSPP and add to the beaker. Dissolve using a magnetic stirrer until solution is achieved.
5c.4. Filter through coarse filter paper to remove any undissolved phosphate glass material.
5c.5. Add 350 ml of the 3.0% dispersant solution into a clean dry multimixer cup.
5c.6. Weigh approximately twenty-five (25) grams of clay to ±0.01 gram.

5c.7. Add the clay to the mixer cup while stirring. Set timer and allow the mixture to agitate on the multimixer for twenty (20) minutes until dispersion is complete.

5d. *Option 1*: Pass the dispersed material through a 325 mesh screen and rinse with distilled water. Tilt the screen and wash all the residue to one section at the edge of the screen.

Transfer the residue to a preweighed evaporating dish by using the stream from the wash bottle. Decant the excess water and oven dry at 220°F to constant weight.

Weigh and calculate:

$$\% \text{ residue} = \frac{\text{weight of residue}}{\text{weight of sample}} \times 100$$

Option 2: Pass the dispersed material through a clean, predried, and preweighed three (3) inch 325 mesh screen. Rinse with distilled water.

Oven dry the screen plus residue at 220°F to constant weight. Determine weight of residue by subtracting the original weight of the screen.

Calculation:

$$\% \text{ residue} = \frac{\text{weight of residue}}{\text{weight of sample}} \times 100$$

5e. *Notes*

5e.1. It is very important that the sample used should be representative of the lot. This is accomplished by appropriate sampling and splitting techniques.

5e.2. It is crucial that the dispersant solution be completely dissolved and filtered. If this is not accomplished, difficult to see +325 mesh TSPP particles can increase the +325 mesh residue.

5e.3. If TSPP stock granules are coarse in grind, it is suggested that a portion (400–500 g) be reduced in particle size through the portable hand grinder and set aside for this test. A smaller particle size will allow the material to dissolve completely and much faster.

5e.4. Sodium metaphosphate or calgon may be substituted for TSPP if necessary.

5e.5. Infrared lamps may be substituted in place of the oven when using the 3 in. screens. Caution must be exercised to ensure that the proper length of drying time is utilized.

6. Test: General Services Administration (GSA) absorption

6a. *Purpose*: This test is designed to evaluate the absorption capacity of granular clays using a petroleum oil or water absorbate utilizing large clay samples for improved accuracy.

6b. *Equipment and materials*

Graduated cylinder: 100 ml.

Glass tube: nine (9) inches in length with a thirty (30) millimeter inside diameter.

Lube oil: AE #10, 30.4° API gravity.

Balance sensitive to 0.1 g.

Clamp stand and clamp.

6c. *Procedure*

6c.1. Pour fifty (50)$\pm$0.1 grams of the material to be tested into the open end of the glass tube. Be careful to catch any fines that may pass through the screened end. Place the fines, if any, back into the top of the tube.

6c.2. Using a clamp stand and clamp, position the glass tube at a 60° angle from horizontal. Place a one hundred (100) milliliters graduated cylinder under the wire end of the tube.

6c.3. Pour seventy-five (75) milliliters of the SAE #10 lube oil into the open end of the glass tube within a fifteen (15) second time period. Mark the time when the first drop passes through the tube and falls into the graduated cylinder.

6c.4. Exactly three (3) hours after the first drop falls, measure the oil in the graduated cylinder and calculate absorption in milliliters per gram as follows:

$$\text{GSA absorption} = \frac{75 - \text{oil in graduated cylinder}}{50}$$

6c.5. The procedure for water absorption is the same with two exceptions:

1. The glass tube is positioned at a 30° angle from horizontal.
2. The waiting period for the dripping is thirty (30) minutes.

6d. *Notes*

6d.1. The percent absorption may be calculated from this procedure by weighing the absorbate and using the following equation:

$$\text{GSA absorption} = \frac{\text{Initial weight of absorbate} - \text{Weight of oil ingraduate cylinder}}{\text{Weight of clay (50 g)}} \times 100$$

7. Test: Total cation exchange capacity

7a. *Purpose*: This test method is used to determine the total cation exchange capacity (CEC) of clay and soil samples.

7b. *Equipment and materials*

Distilling apparatus: boiling flask with cooling condenser.
Condenser flask to absorb ammonium vapors.
Burette: fifty (50) milliliter.
Two (2) buret stands and appropriate clamps.
Erlenmeyer flask.
Buckner funnel or equivalent.
Filter paper.
Glass beads.
Analytical balance.
Mortar and pestle.
Ammonium acetate: 2.0 N (normal) solution.
Sulfuric acid: 0.1 N solution.
Sodium hydroxide: 0.1 N solution.
Sodium hydroxide: 6.0 N solution.
Bromthymal blue solution.
Neutralized methyl alcohol.

7c. *Procedure*

7c.1. With a riffle splitter, separate five (5.0) to seven (7.0) grams of clay from the material to be tested to ensure a representative sample. Grind the separated material with a mortar and pestle to pass a 200 mesh sieve.

7c.2. Weigh approximately one (1) gram of the ground material on the analytical balance and place in an Erlenmeyer flask. (It is advisable to set up the samples in duplicate.)
7c.3. Determine percent moisture on the remainder of the ground material. Calculate dry weight of clay in flask.
7c.4. Add 100 ml of 2 N ammonium acetate to sample of clay in flask. Stopper and shake. Allow sample to soak for at least four (4) hours (24 hours if possible). Shake periodically while soaking.
7c.5. Fit a filter paper to a Buchner funnel, attach to a vacuum line, and wet the paper with 2 N ammonium acetate.
7c.6. Swirl the Erlenmeyer flask so that the clay is dislodged from the bottom and rinse into Buckner funnel. Be sure all the sample is removed from the flask.
7c.7. After all the solution is filtered, slowly wash filtrate with an additional 100 ml of 2 N ammonium acetate.
7c.8. Carefully wash filtrate thoroughly three (3) or four (4) times with 25 ml portions of neutralized methyl alcohol (pH 7.0).
7c.9. Transfer filter paper and sample to distillation flask.
7c.10. Place the discharge end of the condenser into a beaker containing 20.0 ml of 0.1 N sulfuric acid. (End must be submerged to condense the ammonium gas.) Add four (4) drops of bromthymal blue indicator.
7c.11. Add approximately 25 ml of 6 N sodium hydroxide solution onto sample and filter paper in distillation flask. Add four or five glass beads.
7c.12. Connect distillation flask to condenser. Be sure that water is flowing through condenser jacket before applying heat to flask. Bring sample solution to boil and allow to boil for a minimum of five (5) minutes. Remove heat and allow time for all of the ammonium gas to reach the beaker with sulfuric acid and condense.

7d. *Calculations*

$$\text{Total CEC} = \frac{20 - \text{titre}}{\text{Weight of sample}} \times \text{Normality} \times 100$$

CEC is expressed in milliequivalents per 100 grams.

7e. *Notes*
7e.1. It may be necessary to increase the sample size to 3 g if total CEC is very low.
7e.2. Either a hot plate, heating mantle, or bunson burner may be used as a source of heat for the distilling apparatus.

SUBJECT INDEX

www.ingramcontent.com/pod-product-compliance
Lightning Source LLC
LaVergne TN
LVHW011803200926
848484LV00003B/774
9780444517012